New Zealand

Wellington
Christchurch

Chatham Islands (NZ)

Auckland Islands (NZ)

Antipodes Island (NZ)

quarie Island (Aust)

Campbell Island (NZ)

Dumont Durville (Fr)

th Bay

Balleny Islands

Scott Island

Magnetic Pole Area

Cape Adare

Hallett Station (USA – NZ)

TRANS

TERRE ADELIE

VICTORIA LAND

Vanda Station (NZ)

McMurdo Station (USA)

ROSS
SEA

SOUTHERN

ANTARCTIC

Ross Island

Scott Base (NZ)

ostok (USSR)

Ross Ice Shelf

King Edward VII
Land

MARIE BYRD LAND

OCEAN

RANGE

Ellsworth
Highland

SOUTH

Pole Station (USA)

Byrd Station (USA)

AMUNDSEN

POLAR

ole of Inaccessibility

SEA

PLATEAU

Eights
Coast

Filchner Ice Shelf

Eights Station (USA)

Ellsworth Station (USA)

General Belgrano (Arg)

ALEXANDER
LAND

ING MAUD
ND

COATS LAND

Halley Bay Station (UK)

BELLINGSHAUSEN
SEA

uin (Belg)

ev (USSR)

BRITISH ARGENTINE CHILEAN BASES

Sanae (S. Africa)

WEDDELL SEA

Maudheim (Nor)

South Shetland Is

Graham Land Peninsula

DRAKE PASSAGE

Cape Horn

h Sandwich Islands

South Orkney Islands

Falkland Islands

S O U T H

Man and Nature in Antarctica

SOUTH

MAN AND NATURE

Text: Graham Billing

Illustrations Editor: Guy Mannering

A New Zealand view

IN ANTARCTICA

A. H. & A. W. REED

WELLINGTON – SYDNEY – LONDON

First published 1964

Reprinted 1965
Revised edition 1969
Reprinted 1971, 1973

A. H. & A. W. REED LTD

182 Wakefield Street, Wellington
51 Whiting Street, Artarmon, Sydney
11 Southampton Row, London WC1B 5HA
also
29 Dacre Street, Auckland
165 Cashel Street, Christchurch

© 1964

Department of Scientific and
Industrial Research, New Zealand

ISBN 0 589 00332 1

Published in co-operation with
the Antarctic Division
Department of Scientific and Industrial
Research, New Zealand

KYODO PRINTING COMPANY LIMITED, TOKYO, JAPAN

PREFACE

ANTARCTICA IS FULL OF CONTRADICTIONS. Its face is emptily inhuman, yet from its silences is coming a rich store of knowledge for scientific research. It is a land of warring elements where men have lost their lives in high endeavour; yet it is a continent of peace where no shot has been fired in anger and no blood shed in war. Antarctic men struggle with nature's forces in order to live and win knowledge for the benefit of mankind; yet under the Antarctic Treaty, the scientists and support-men of twelve nations work in harmony and mutual trust, demonstrating to the world a successful experiment in international cooperation.

Since 1955 New Zealand has been an active participant in this exciting paradox and it has been my privilege to have been associated with New Zealand Antarctic endeavour since its inception. For long it had been my ambition to tell the world about the Antarctic and New Zealand's part in it, but it was not until I had the good fortune to recruit to the New Zealand Antarctic Research Programme Graham Billing, author, and Guy Mannering, illustrations editor and photographer, that I could see the fulfilment of my ambition.

Before they went to the Antarctic in October 1962, I was able to sow in their minds the ideas for this book. A luxuriant harvest has now been reaped in the production of *South*. Whilst Billing and Mannering have been the inspiring minds in its production, the many Antarctic men who have contributed illustrations and have assisted the author with the scientific complexities of Antarctic research have a share in this book. It tells an exciting story and portrays a dramatic picture of New Zealand endeavour in the Antarctic continent.

G. W. Markham

Superintendent, Antarctic Division,
Department of Scientific and Industrial Research
Wellington, New Zealand. 1964.

CONTENTS

SOUTH

Ross Island has been a base point for Antarctic exploration since the first journeys into the interior were made by Captain Robert Falcon Scott's *Discovery* expedition of 1901–04. McMurdo Sound is the sea arm between the island and the mainland coast, part of the deep Ross Sea embayment where, in summer, open water allows the closest approach to the Pole by ships.

At the head of the Sound the United States Navy has concentrated the logistic effort of Operation Deep Freeze which supports its scientific stations on the Polar Plateau. New Zealanders at Scott Base live only two miles from McMurdo Station with its nuclear reactor, oil tank farms, heliport, machine and aircraft shops, church and storage buildings so that the two nations associate in the full spirit of the Antarctic Treaty.

Yet Antarctic weather can make those two miles as long as two hundred. Turned southward to the Ice Shelf, Scott Base retains an air of isolation. Far out on the ice a dog-team party homeward bound sees only the stark outline of the peninsula and perhaps the Scott Base aircraft hangar thrust up by the mirage into a bright green tower. American aircraft flying from Williams Field at the head of the Sound look no bigger than birds appear in the smaller scale of temperate lands and mountains. From the Ice Shelf, Mt Erebus appears a real giant; the 13,000 foot peaks of the Royal Society Range tower over the Sound from the western side; and thirty miles away an iceberg, magnified by the mirage into a castle keep, guards the entrance to the sea.

In days of calm it appears a changeless world. There is no obvious sign of the enormous forces working on the continental land mass, no sign of the life which is going on in the sea below the ice shelf, in rock crevices on the mountain slopes and in the cold, clear air itself.

Be careful; for here you are walking in a land where life and non-life, physical movement or motionlessness, are in delicate balance. On that hillside you may pick up a rock, breathe on the tiny, frozen insects underneath it and so bring them to life. Sledge across that sea ice and it may relent under the sea's pressure, part and drift away with you as a floating island.

Boundless energy is stored here although the balance of forces gives a false impression of a land relaxed and still. The Antarctic ice mass is a giant engine of cold, whose influence extends about the whole Southern Hemisphere from convulsions of winds in the stratosphere to the irrepressible currents of the world's oceans.

Here the Earth's magnetism is concentrated round the South Magnetic Pole, bending the molecular particle emissions of Sun and Cosmos into an eccentric halo which glows as the winter night's aurora. Here is contained ninety per cent of the world's ice with power to raise the oceans some 200 feet and pressing the crust of earth itself more than 2,000 feet towards the core. Here are the world's lowest temperatures and strongest winds, hardiest animals and seas richest in life. Here is Antarctica, a continent bigger than the United States and Europe combined, averaging

10

more than 9,000 feet in height, almost uninhabited, barely known and holding many secrets sought for by man in his twentieth century search for the prime sources of physical energy.

What causes a solar flare? By what processes does the sun generate its vast energy and radiate it out to our Universe? What causes an earthquake and why does it appear that earthquakes do not occur in the Antarctic land mass? Beneath the rock crust of the Earth lies an unknown core. How do the microscopic vegetable organisms of the Antarctic sea turn sunlight and chemicals into such profuse life?

These are among the problems before twentieth century science seeking out the sources of energy with which to build new nations, new environments, new foods, to reach out into the infinities of space, to control life and death.

The search is the task of basic scientific inquiry; of geophysics, which embraces all aspects of the physical universe; of geology, which seeks to discover how and of what the Earth is made; and of the sciences which seek to find the basis of life in man and nature. As an educated nation, New Zealand has been inevitably drawn into the search and into Antarctica, the world's great laboratory for basic science.

The Antarctic continent is now defined. Its heights and depths, breadth and configuration, its knowns and unknowns, have been assigned values during 150 years of exploration. The new Antarctic age is a creative one in which man, having examined the continent, is becoming its user, putting his intellect to work in exploiting its resources. Antarctica is an undeveloped country and creative thought and action are realising its assets.

New Zealand's nearness to the Ross Sea side of Antarctica (McMurdo Sound is 2,400 miles south of the capital city, Wellington) has brought involvement in Antarctic exploration since Cook's first circumnavigation of the Southern Ocean and Sir James Clark Ross's astonishing voyage of discovery in the early 1840's. Ross sailed on his second voyage from the Bay of Islands in New Zealand's far north, Scott and Shackleton made Lyttelton their springboard when they began to open up the continent in this century.

It was on the basis of a British claim to Antarctic territory, the Ross Dependency, a wedge-shaped segment bounded by the South Pole, latitude 60°S and longitude 160°E to 150°W, that New Zealand became geographically involved. Britain handed over the Dependency to New Zealand in 1923.

The territorial claims of nations working in Antarctica are now "frozen" under the Antarctic Treaty, signed in 1959 and finally ratified by all twelve powers in 1961. New Zealand played a leading part in the Treaty's formulation, being unreservedly willing to abandon territorial claims in favour of an international regimen for the continent.

Consequently her voice in Treaty consultative committee meetings is a respected one, for the nations know New Zealand has no territorial bias and thinks selflessly of international cooperation. At meetings of the Scientific Committee on

11

Antarctic Research (S.C.A.R.), a committee of the International Council of Scientific Unions set up to facilitate the exchange of scientific information and to co-ordinate research, her leading scientists make valued contributions chairing working groups for individual disciplines and collaborating in general discussions.

The Department of Scientific and Industrial Research (D.S.I.R.), now organises the New Zealand Antarctic Research Programme (N.Z.A.R.P.) through its Antarctic Division. Advising and assisting the Antarctic Division in programme policy-making and liaison through S.C.A.R. with the programmes of other nations is the Ross Dependency Research Committee (R.D.R.C.), set up in 1958. The R.D.R.C. is also the Royal Society of New Zealand's National Committee for Antarctic Research, for the Society has been one of the principal forces in the institution and extension of N.Z.A.R.P.

With its small staff and offices in busy down-town Wellington, the Antarctic Division is more subject to seasonal bursts of activity than any other Government Department. While running an Antarctic programme is a complex year-round business the task of mounting an expedition means a long period of intense activity each spring.

Expedition members have to be briefed and trained in polar living on the snow slopes of the North Island volcano, Mt Ruapehu, outfitted with special clothing and shepherded to the ships and aircraft which take them south from Lyttelton or Christchurch. Stores and equipment must be purchased and packed. Daily radio conferences between New Zealand and Scott Base are held to insure a smooth operation. By the time the first United States Navy and Royal New Zealand Air Force aircraft fly south in October most of the base relief arrangements must be complete.

New Zealanders consider Antarctic work among the most rewarding fields of adventure open to them. When applications are called each autumn for the thirty-five or so expedition jobs, hundreds of letters jam the Antarctic Division mailbox.

The Division's Superintendent and the Scott Base Leader for the new season travel from one end of New Zealand to the other interviewing applicants, searching for a score of men whose minds and constitutions, as well as their professional qualifications, fit them for living through the winter night at Scott Base or Vanda Station, and for others whose alpine skill will protect them through the summer field survey and geological expeditions by motor toboggan or dog sledge. Often they see the same faces, for once a man has lived in Antarctica he longs to go again.

At the same time, other D.S.I.R. divisions, the universities and mountain clubs are selecting men to carry out special research projects during the summer apart from those continuing programmes organised by Antarctic Division. The Victoria University of Wellington Antarctic Expeditions in geology and geochemistry, and the Antarctic Biological Units of Canterbury and Otago Universities are well established elements of N.Z.A.R.P., and all are mounted under the

12

ultimate control of D.S.I.R.

During the winter the Royal New Zealand Navy also refits its Antarctic support ships, *H.M.N.Z.S. *Endeavour* for another tough summer of ice navigation in which her crew make voyages south carrying Scott Base stores and men as well as petroleum fuels for both New Zealand and United States bases.

These services together with free use of the International Airport at Christchurch, the Port of Lyttelton and the provision of base facilities at Christchurch are given to the United States by New Zealand in return for air support, both transoceanic and on the continent, and other logistic assistance.

Along with all Antarctic research New Zealand's scientific inquiries in Antarctica can be divided into four different categories. Unique conditions of any sort must be studied. If a particular experiment can be carried out to greatest advantage under Antarctic conditions or in Antarctic latitudes it must be done there. If an inquiry requires a series of observations on a global scale before all relevant data can be collected, Antarctica must have observatories. Lastly, the ordinary processes of discovery, exploration and the compiling of geophysical, geological, biological and meteorological information about things as they are must continually be pursued. In the light of the new Antarctic science the pursuit seems as endless as the Universe itself.

Scott Base is the pivot point of these investigations, servicing permanent and temporary satellite bases and expeditions within a radius of several hundred miles. The furthest outpost is at Cape Hallett 350 miles to the north where biologists study penguin and other seabird life during the summer. Hallett Station was once a key research base. It was founded as a joint United States-New Zealand station for the I.G.Y. but abandoned as a wintering-over station after fire destroyed much of the scientific laboratory space in 1964. The station was a key staging point for New Zealand expeditions exploring north Victoria Land.

On Ross Island New Zealand maintained a summer biological station at Cape Royds for several years but its role has now been taken over by Harrison Station at Cape Bird further north, where a larger penguin colony than the one at Cape Royds offers broader research opportunities.

During the southern summer of 1967 a major advance in New Zealand Antarctic science was made with the construction of Vanda Station, the nation's first permanent base on the continent itself. It was an operation unequalled since the construction of Scott Base ten years earlier. R.N.Z.A.F. Hercules transport aircraft dropped building materials and stores. Tractor trains and Operation Deepfreeze helicopters took men, provisions and scientific equipment across the McMurdo Sound sea ice, over the continental ice and rocky valleys. Thus one of Antarctica's most important research stations was established by New Zealand and opened as an international refuge at which scientists of all nations working in the unique snow-free region are welcomed.

* Withdrawn from Antarctic service in 1971.

2. *The Elements*

COLD HAS NO SOUND, NO SMELL. COLD CAN WOUND AND KILL, depress or exhilarate the spirit, crush a continent, subdue oceans, cleanse the air and make the land sterile. Cold has no substance yet in the mind it is the most substantial of Antarctica's elements.

It grants power to walk across the sea but makes fire a great danger because it leaves no water. It creates the wind, the mirage, the round rainbow of a sun halo. By its influence a man can stand amazed at the new, clear wonder of the night sky.

Exposed to a temperature of minus 22°F in a twenty-two mile-an-hour wind inactive human flesh will freeze in one minute. A man who falls into sea water half a degree warmer than freezing will die in fifteen minutes or less. Flesh feels no touch or pain when skin temperature falls to 50°F. Muscles become weak and blood stops flowing in the skin and limbs. When the core temperature of his body falls to 79°F a man enters a state of coma leading into death.

Cold wounds a man's flesh with frostbite when ice crystals form in the cells and tissues, dehydrating protoplasm, the basic substance of life, and killing tissue by preventing the circulation of blood with its supplies of oxygen that maintain life. In the painful hours of thawing out frostbitten flesh swells and later the skin will crack and blister. After the skin has healed and smoothed, scars remain under the surface, ready to freeze first when the cold attacks again.

Frostbite unites Antarctic travellers in a common bond of vigilance because each depends upon his companions for warning if a nose, chin or cheekbone takes on the dead-white pallor of a cold injury. Caught quickly, the bite can be thawed by the warmth of a mittened palm. Frozen fingers go white like candle stubs.

Antarctic sledge dogs stop panting when they lie still in the cold, and so lose little heat through their lungs. In blizzard winds they curl up in a ball to expose

14

less skin and to let the snow cover and protect them. Seals have blubber to insulate their bodies and on cold, harsh days can sleep in the sea instead of on the ice. Penguins' feet have few blood cells; they sleep rocking on a tripod of heels and stiff tail feathers.

But without any of these advantages of adaptation men must confine themselves in clothing and heated buildings. Antarctic life is a constant putting on and taking off of clothes, opening and closing of double doors.

Animals keep warm by erecting their body hairs to increase the thickness of still and non-conductive air which covers their skin. Man substititutes clothing, whose effectiveness depends on the temperature difference between its inner and outer surfaces. Antarctica is a laboratory for research in the clothing and textiles which keep men warm, and the ingenuity applied to this research has made long periods of healthy and productive outdoor living possible on the continent.

Since the body's rate of heat loss rises steeply with increasing wind velocity, Antarctic clothing is designed on three principles. It must have a windproof outer surface, although waterproofing is unnecessary: it must trap a thick layer of air: and it must prevent movement of that air over the skin. A colourful windproof anarak with a fur-edged hood and windproof trousers with deep pockets fulfil the first condition; woollens and an eiderdown jacket or vest, the second; and a string singlet, with its mesh of air-trapping cells, the third. Within this range the individual's choice of outer clothing can be varied to suit the temperature.

Feet are protected by mukluks, big fabric and rubber boots with thick insoles. Heavy woollen socks are worn inside the bright yellow quilted nylon uppers.

Antarctic huts provide a similar sort of protection from the elements on principles of insulation related to those used in clothing design.

Scott Base was designed by New Zealand's Ministry of Works as a series of self-contained, heated units connected by a covered walkway. Even in a hundred-knot blizzard the men can move from one hut to another in still air. Because the huts lie at right angles to the southerly blizzard wind, and spaced well apart, fire is less likely to spread from one end to the other.

Each hut is like an inside-out refrigerator with two heavy, narrow doors. Door handles can be unlatched with a push and are well covered with fabric because bare hands can stick tight to frozen metal. Roofs are flat, so that blizzard-blown drift snow does not settle but passes straight over them. Ceilings are criss-crossed with heating ducts that pour warm air into every corner when the thermostatically controlled heaters are running. Windows are tinted against snow glare and double glazed.

Even indoors cold cannot be forgotten. There is ice or frozen earth under each hut. A can of beer on the floor will freeze solid while ceiling temperatures are tropical. Men in top bunks can sleep with one blanket while those on the bottom must creep under five.

15

SOUTH

Only in brief mid-summer does the cold relax at McMurdo Sound and around Antarctica's shores. The dark rocks absorb heat during twenty-four hours of sunlight. Snow melts and melt-water streams rush down roads, gullies and the Scott Base covered way. Dust blown from the hillside, mud dropped from vehicle tracks, dog droppings, dumped rubbish and scraps of seal skin and blubber absorb the Sun's warmth to sink slowly through the snow and ice leaving an untidy, soiled landscape.

Men try swimming in melt-water pools or strip to the waist to work in a sheltered corner. They build "sun-coffins" lined with aluminium foil to achieve an "all-over" tan, or simply lie blissful and at ease among the warm rocks of a brown, coastal hill.

In this short season a keg of beer shipped from New Zealand may be enjoyed out of doors by men in shirtsleeves although it is muddy underfoot. These are the hey-days of Antarctic living, the weeks from Christmas to early February, when New Year greetings are made to a golden sun which has never set.

Snow evaporates under the benign glow. Lichens flourish, insects hop among the rocks, penguin chicks grow fat and skua gulls raucous. Seal pups leave their mothers. Sledging men ski in their underpants. Ice-cream fails to freeze in the base's covered way.

But the influence of cold is still here. Tobacco turns crumbly in the dry atmosphere. Ice blocks must still be cut daily to fill the melters which supply the base with water. High winds bring less blowing snow because the loose surface particles have melted and coagulated, but their cold bite is still keen. Sewage must still be heated before it can flow down insulated pipes to the tide crack where the sea ice joins the land.

Antarctica's year begins in these high summer weeks. Ships can close the land, stores are unloaded and checked, field parties move more freely with their motor toboggans among the mountains of Victoria Land, new huts can be built and scientific equipment is overhauled in comfort.

Autumn comes quickly and summer support staff leave the base in late February when there is already a hint of darkness, for an hour or two at midnight. Winter party men reach for their eiderdown clothing more frequently and soon will wear it all the time. Temperatures will be at their lowest in July and August after the Sun has begun to return and the greatest effect of the winter-long cooling is felt.

Scott Base then records its lowest temperature for the year—at least ninety degrees of frost, compared to which the midsummer zero seems a marvellous, an unbelievable warmth. By early October when the first relief staff arrive by air from New Zealand temperatures are ranging between fifty and seventy degrees of frost.

Newcomers feel as if they are shrivelling, but the winter party men make casual remarks about the heat. They have become acclimatised, their bodies have learned to maintain heat balance under Antarctic conditions, producing more heat by chemical processes and expending less energy on shivering.

The new season's men soon become acclimatised too and find that work in the cold need not be unpleasant except in extreme conditions of winter darkness or wind. A man can sweat profusely, even while the air is stinging his nostrils until they feel as if they are full of soft needles of ice, while his beard goes white with rime and ice and his eyelashes grow thick with rime, sticking together when he blinks.

Every field worker has his own ideas on how to protect his face from wind-chill—a cloth bandage nose-guard, a face mask to cover nose and mouth, a clear plastic blizzard visor like an enormous, broad bird-bill which stops drift snow from coating the face, and so thawing and re-freezing into a mask of ice. A beard, too, is helpful although it easily becomes matted with ice.

In the colourful language of the meteorologist, Antarctica is the world's greatest heat sink. It is the coldest part of the earth which, though symmetrical in shape. is asymmetrical in climate. Where the Arctic is an ocean surrounded by land masses, Antarctica is a land mass surrounded by ocean. The winter growth of Antarctica's pack ice greatly increases its effective area as a producer of cold and a consumer of heat from warmer regions.

Antarctica's height alone makes it on average more than 30°F colder than the Arctic, and the rest of the large difference in average annual temperature occurs because the Antarctic ice cap reflects heat while the Arctic Ocean tends to absorb it. The South Pole area is so sunny that about eighty per cent of radiated Sun heat reaches the snow though most of it bounces off again.

Air circulates about the Earth because the Sun heats some parts of the atmosphere more than others causing differences in pressure which must be equalised by air movement. That is why Antarctica, the heat sink, is so important in the world's weather system.

Antarctica, like any other part of the Earth, must exist in a state of heat balance. Warm, moist air flowing into the Antarctic region also provides heat directly and through the release of energy when its moisture freezes out. Thus the continent's loss of heat through re-radiation of heat from the Sun is partly compensated.

Antarctic mirages exist because the radiation of heat causes a temperature inversion in the first few thousand feet of the atmosphere—the air becomes warmer instead of cooler with increasing altitude. Air layers of different temperatures have different densities and so refract light rays to form the distortions of the mirage, the most spectacular of which is the *fata morgana*, a wavering mirror image upside-down on top of its reality.

Antarctica's exchange of heat with the rest of the world involves a complex system of great and small movements in the meteorologist's three regions of the air: the atmosphere, troposphere and stratosphere. Stratospheric winds whirl in winter about a cyclonic vortex centred over the Polar Plateau. In the atmosphere beneath this vortex, great cyclonic storms roll round Antarctica's perimeter bringing snow to the coasts.

This is one sort of Antarctic storm. The other, far more terrible, is caused by the layer of cold and heavy air which lies across the ice sheet as a vast and almost permanent anticyclone. Because it is heavy, the air in this system simply falls down the Polar Plateau ice slopes at speeds which, at the coast, may exceed a hundred miles an hour. Known as katabatic winds these air movements can bring the blizzard in which transported snow buries men and their possessions in a swirling drift but on the Plateau itself they average thirty to thirty-five knots.

The passage of the katabatic wind across the snow carves ripples and hummocks called sastrugi and snow dunes sometimes several feet high. Travellers retracing their steps on the Plateau may find their tracks weeks later after loose snow has been blown away from the edge of each packed imprint.

Antarctica's annual rate of precipitation is equivalent to about six inches of water. In theory it would take 10,000 years for the present amount of ice to accumulate if this rate remained constant. Much of each year's snowfall, however, is blown away by the katabatic wind, and some evaporates. Snow accumulates on the surface by under six inches a year on the average with less on the Plateau and more on the coast. Because the ice flows outwards under its own weight until it breaks off into the sea as icebergs the continent's altitude does not increase at the same annual rate. The present annual rate of mass increase for the ice cap is thought to be very slight.

Falling snow is not a common occurrence at Scott Base. In mid-summer fresh snow on the rocks, bare ground and dry ice soon evaporates and the blizzard drift snow which blankets the spring hillsides of McMurdo Sound also disappears during summer. While rain has not been recorded at McMurdo Sound it falls occasionally on the northerly coasts and in the Dry Valley area.

Ice fog also adds minutely to snow levels. When warm moist air flows in from the sea and is quickly cooled, tiny ice crystals form and fall. Hoar flowers grow on warmer, cloudy days covering smooth ice surfaces with fragile lace petals, so thin and delicate that they shiver in a light wind.

Cloud also brings whiteout, a condition hazardous to travellers and dreaded by aircraft pilots. Light in a whiteout is so perfectly reflected between cloud and snow that there are no shadows, no horizon, no surface definition, no marks by which to gauge height or depth. Walking in a whiteout is like trying to run down a broken flight of steps in the dark.

Antarctic temperatures are not standard but can vary widely over a short distance. Scott Base is several degrees colder than McMurdo Station's sunny corner on the northern side of the Hut Point Peninsula. Its average temperature for a year is about minus 26°F or forty-eight degrees of frost, and with its winter minimum of about ninety degrees of frost it is very much warmer than the Russian-

18

recorded minus 126.9°F at Vostok Station, one of the most inaccessible on the Plateau.

The difference in average temperature between the South Pole and a coastal station near the Antarctic Circle (66°32'S), may be as great as, or greater than that between a Florida beach resort and the Canadian Arctic. A northern European town may experience temperatures just as cold as those of a spring day at the edge of the Ross Ice Shelf.

Another aspect of Antarctic climate which is now assuming importance is the study of the micro-climates which surround living things. Minimal changes in temperature and humidity greatly affect the delicate balance between existence and non-existence in a harsh world.

Most of man's knowledge of Antarctica's elements and how they affect the world's weather has been gained since 1957 and the beginning of the I.G.Y. The continental climate is unique, incompletely known and understood but of great importance to the rest of the world as are the influences of the little-known circumpolar oceans.

At Scott Base meteorological work is limited to daily recording of local surface weather conditions because the United States at McMurdo Station carries out a full programme of observation and research. During and after the I.G.Y. several aspects of solar radiation and surface temperature fluctuations were investigated at Scott Base, and equipment for filtering from the air dust-particles which form the nucleus for atmospheric ice crystals was operated for the Australian Commonwealth Scientific and Industrial Research Organisation. New Zealand's meteorological research is now concentrated at Vanda Station. The Dry Valley area is particularly important for Antarctic weather studies. Rain sometimes falls there but very little snow. Glacial ice has retreated from the valleys perhaps partly because of a change in local climate. The lakes of the region have ice sheets which thin at the edges in summer when melt water streams run into them. Soils are forming under sparsely known conditions of weather erosion. The lakes have warm bottom water, a condition certainly related to the anomalous climate, and both lakes and rock weathering provide a record of past climatological change unique on the continent. Meteorological research is also vital to the general study of how the snow-free area was created and why it remains.

3. *The Ice*

DOME-SHAPED AND ROUGHLY CIRCULAR, THE ANTARCTIC ICE sheet is proportioned to appear like a weight which keeps the planet Earth perpendicular. It is the parent and source of all Antarctica's manifestations of ice, of the glaciers which carve their way seaward through mountains, of the ice shelves which stretch like flat white deserts across coastal bays, of the sea ice which forms each autumn, and of the snow itself.

Ice moves, flows, changes, diminishes and grows. The ice sheet continually replenishes itself by causing snow to fall and under this slow accumulation spreads steadily outwards like a domed jelly subsiding under the pressure of its own weight. A snow crystal which fell at the Pole of Inaccessibility would take tens of thousands of years to reach the coast with the steady radial flow of the sheet, but it would eventually enter the sea and melt again. It might move down a broad glacier or down a slope where the sheet itself flows clear to the continental edge.

About a million years ago the Earth began to grow colder. Perhaps, in a moment of cosmic proportions, its spinning flight about the Sun was checked or altered. The continent of Antarctica, like Greenland in the polar north, became incarcerated in ice which waxed and waned four times across the Earth's surface. Now the slow tide of the ice is receding as the Earth warms again, but as long as the planet continues spinning on its present axis Antarctica's ice is likely to remain. It creates its own climate and is the victim of it.

The ice began among the mountains of East and West Antarctica, and areas east and west of the Greenwich Meridian. Slowly it filled mountain cirques with snowfields which spread out and lapped down to plateaux and lowland plains. As the ice grew its weight depressed the continental lands and islands, forcing their solid rock structure deep into the plastic rock of the Earth's mantle far below the crust. The sea ran in to fill the low coastal valleys of a once green land; froze, and added to the sheet. At last the ice sheets of East and West Antarctica converged to form their present mass.

Now the warming Earth is wearing slowly at the ice sheet's edge and yet the same warmth brings growth and spreading flow to the surface of the sheet because the Antarctic atmosphere is fed with warmer air carrying more moisture for snow making. Thus Antarctica endures an uneasy state of paradox in which the same forces both diminish and increase it.

When ice retreats along a coastal edge the land responds by rising under released pressure. In this way much of the mere two per cent of the Antarctic land visible above the ice has become free. Man can now look closely at the complex relations between air, land, sea and the ice. Such study together with investigation of the behaviour of the ice itself is glaciology, a science of growing importance to New Zealand's Antarctic Research Programme and to the work of all Treaty nations.

In Antarctica the ice sheet embodies all the authority of nature dominating land and life in a way unknown to any single source in other latitudes. To live in Antarctica is to learn to use the ice and not to fear it, to treat it with respect but to adapt to the environment it provides without attributing to it malevolence or treachery. To live on the ice, as on the sea, demands resourcefulness, patient observation of its behaviour, adaptability. Like the sea, the ice gives no second chance but is entirely neutral in the struggle for survival.

Ice holds the impartial power of chance life or chance death over Antarctica's animals. Adelie penguins starve if the sea ice beside their breeding rookeries does not break up in summer, yet ice is their refuge from the animal predators of the sea in which they live. Emperor penguins lay their eggs in winter on the sea ice about coastal islands, rear their chicks on it in early spring and then, when it breaks away from the land, use it for safe transport to the rich feeding grounds of the pack ice belt further north. Seals die when their teeth are worn to stumps from gnawing breathing holes through the ice but they too bear their young on it and escape to it from enemies in the sea.

In the same way the ice dominates human life in the Antarctic and amazes human sight with the brilliance of its surfaces and the violet, emerald and cobalt depths of its shadowed faces.

The ice environment has six separate aspects for men in Antarctica. They are the ice sheet of the Polar Plateau, the glaciers which flow off the Plateau, the ice shelves, the fast sea ice, the pack ice and icebergs. All have their separate dangers, their separate behaviour and their beauty, separate or in conjunction with each other or the bare and many-coloured rock of mountain faces.

At the South Pole on a summer's day the air is likely to be clear and still. The cold will be intense for here you are ten thousand feet high as well as near the centre of the coldest continent. Stand at the South Pole itself, turn in a circle and you will see the Polar Plateau ice sheet, a level plain in these parts, stretching out in all directions. The sastrugi give it the appearance of heavily grained wood, white and shining.

You stand on snow, but if you dug below the surface as the Americans have done to build their South Pole Scott-Amundsen Station you would find the snow transforming slowly under mounting pressure into firn or coarse-grained snow and then at about 300 feet, pressed into ice.

The continent itself lies about 7,500 feet beneath you. The sheet on which you stand contains about 7,200,000 cubic miles of ice, is 2,800 miles in diameter and more than 15,000 miles in marginal length forming ninety per cent of the world's ice. It traps two per cent of the world's total water supply or about 6,450,000 cubic miles and, should it melt, would raise the world's oceans by at least 200 feet.

Not everywhere is the ice sheet as level as at the South Pole. Its upper surface reflects the shape of the long-buried bottom and in the immensity of the white Plateau are great steps and terraces covering frozen mountains. Slowly rolling dunes of ice, strained and deeply crevassed, cover the steppes and grasslands of a continent which once was warm.

The sheet reaches its greatest height about the Pole of Inaccessiblity where Russian explorers judged the height at some 14,000 feet. Both height and distance from the sea give this point a mean annual temperature of minus 71°F.

Despite its vast mass the ice sheet moves but the mechanics of its movement is still a matter for conjecture by glaciologists. Ice combines the properties of both viscous and plastic conditions and seems to move by sliding over the land, perhaps melting where it touches rock under pressure, and by a spreading flow within its mass.

It is thought that deep in the sheet the individual ice crystals change their shape or re-orientate their positions under pressure until the mass is in the right condition for flow to take place.

Crevasses and ice falls form where the ice flows over irregularities in the land and is strained beyond breaking point. Sometimes the crevasses are hidden by snow bridges and are a great danger to travelling field parties with dogs or vehicles. In places where the katabatic wind has swept the blue ice bare of snow the crevasses loom like huge and shadowed gashes in the slopes. Some grow up to 100 feet wide and 300 feet deep while others remain mere disconcerting cracks in the surface.

One of glaciology's most pressing problems is to find out if the ice sheet is gaining or losing mass, a question closely related to long-term climatic trends elsewhere in the world. At present there are indications that the accumulation of snow on the surface exceeds in volume the wastage into the sea at the edges from icebergs or wind blown snow. This means that the sheet is at least growing thicker. Snow, like sand, is a protean substance, however. It might be present today and wind-carried fifty miles away tomorrow. Precipitation of snow cannot be accurately judged and net accumulation for a particular area must be gauged instead.

Stakes planted in the snow surface can tell this change, and if the glaciologist digs a pit he can expose summer and winter snow layers for succeeding years. Seasonal layers have different textures because they were laid down under different temperature conditions and so annual snow accumulation can be read on the pit face.

The temperature at about fifty feet is equivalent to the mean annual air temperature on the surface. Because the layers vary in temperature they also contain different proportions of isotopes collected in the atmosphere in the precipitation process. These are the atmospheric isotopes of oxygen and hydrogen, and because their concentration varies with temperature and depth they also provide a means of separating and dating the layers. A snow mine gives a yearly record of the atmospheric past in such a way that concentrations of radioactive debris from atomic and nuclear bombs can be discovered and conclusions drawn about circulation of air at various levels of the atmosphere.

The ice sheet moves and spills its bulk through the channels it has carved, the rock-walled valley glaciers most common to Victoria Land where New Zealand Antarctic research is concentrated.

The glaciers are routes for Plateau exploration, staircases up the sheer alpine wall which stretches from Cape Adare in the north of the Ross Dependency and close to the Antarctic Circle to within 300 miles of the South Pole. Some are impassable by any men or vehicles; others negotiable only with difficulty by skilled men with dog teams or motor toboggans, and others are broad highways to the Pole though still dangerously crevassed in places. All have cut away the rock to expose the strata of which the continental land is built.

In Victoria Land the Antarctic valley glaciers attain their most perfect development, feeding ice into the Ross Ice Shelf in the south and directly to the sea through floating ice tongues north of Ross Island. Their baring of the land has meant that about one third of Antarctica's exposed rock lies within the Ross Dependency. Glaciologists and geologists as well as other geoscientists whose disciplines investigate the nature of the land beneath the ice and the air above it work here easily in the co-operation demanded by a continent in which geophysical problems are more closely related than in any other part of the earth.

There is evidence that Antarctica's glaciers, along with those of the rest of the world, are shrinking. The valley glaciers flowing into McMurdo Sound are thought to have once been about 2,000 feet thicker than they are now and some have receded right back to the plateau edge leaving the ice-free area known as the Dry Valleys. Released from the glaciers' weight the McMurdo Sound coast has risen again and bears marine fossil evidence of its time under the sea.

Probably, the recession began about 10,000 years ago and is now coming to a close. The ice sheet responds sluggishly to temperature changes and if its present postulated thickening is indeed occurring it will be a long time before it responds by flowing outwards again. The paradox of warming climate, marginal retreat but increasing snowfall again makes itself apparent and its explanation remains indefinite. The only certain fact is that a very large upward temperature change would be necessary to overcome the heat-radiating action of the present ice volume and allow the ice to melt.

A thousand feet thick at their seaward edges and thickening towards the land, floating ice shelves front one third of Antarctica's coast line. The biggest is the Ross Ice Shelf, 200,000 square miles in extent and 400 miles wide, filling the southern triangle of the deep Ross Sea embayment. The Ross Ice Shelf is fed by the glacier streams from the ice sheet and the heavy snowfall on its own surface. It spreads northward flowing under its own weight about a thousand yards a year and calves, or releases to the sea, tabular icebergs sometimes small and sometimes of a hundred square miles or more.

The embayment filled by the Ross Ice Shelf is balanced on the other side of the continent by the Filchner Ice Shelf which fills the smaller southern triangle of the Weddell Sea south of the Atlantic Ocean. Other shelves are much smaller but their behaviour is similar and responsible for several very complex problems for glaciologists.

Though ice shelves calve icebergs along a ragged edge, their shape remains roughly constant and related to the land features around and underneath them. Made of snow, firm and solid ice the shelves even rise and fall with the slight tides and have a "strand crack" where they join the land. Snowfall is thought to be the most important contribution to their growth.

The ice shelves have been studied since Antarctic exploration became established in this century, but their science is not yet far advanced. The strains which a shelf undergoes before a piece breaks off to form an iceberg, and the manner in which it flows towards the sea are not yet well known. Shelves may melt on the bottom or they may grow downwards by freezing the sea.

Oceanography becomes involved with glaciology in the consideration of this unknown for sea currents and temperatures must affect the shelf. Biology joins the study when questions are asked about the possibility of life existing under a permanent ice shelf. Phyto-plankton, the vegetable growth in the sea on which oceanic life cycles are based, needs sunlight to conduct photo-synthesis, the process of converting nutrient chemicals in the sea into living matter. There is no sunlight under the ice shelves. Biologists, oceanographers and glaciologists together ponder the possibility that a curiously adapted life cycle may be going on underneath the shelves.

New Zealand's Antarctic glaciological studies are now carried out by scientists and surveyors of Antarctic Division. These studies have been concentrated where the

24

ice meets the sea close to Scott Base giving an opportunity clearly to define the relationships between shelf, sea and the sea ice newly formed each year. The study of shelf ice is being carried out on the locally-named McMurdo Ice Shelf, an off-shoot of the Ross Ice Shelf bounded by the south coast of Ross Island, White Island, Minna Bluff, Black Island and the Brown Island Peninsula. This small shelf gains sufficient mass from the glaciers which flow off Ross Island to give it localised behaviour, and make it an ideal model for research work.

The McMurdo Shelf is small enough to be completely encompassed by every type of measurement and observation, from estimates of its thickness to the recording of its rate of flow in a particular direction. When this shelf is fully known many of the answers to the chemical composition, behaviour and mass budget of its vast relations will be provided.

Some measurements of the McMurdo Shelf were made during the I.G.Y. but a thorough long term research programme was not begun until the 1962–63 summer. On the surrounding landmarks, beacons were erected to give an accurate survey network on which to calculate the movement of marks on the shelf itself. Bright orange movement markers were set up at various points so that whichever way the surface strained and moved the distance could be measured. Other poles recorded surface accumulation of snow or its opposite, ablation, at points within a radius of fifty miles from Scott Base.

Various methods are being used to calculate the thickness of the ice at different points, among them radar by which a radio signal is reflected off the water-ice surface and the sea bottom so that the distance it has travelled and thus the ice thickness can be measured. The problem of whether the shelf is freezing deeper at the bottom or melting is being investigated by similar methods.

Before this study the chemical properties of shelf ice were largely unknown. There was surprising evidence of the presence of brine, presumably from the sea, trapped or circulating deep in the core of the shelf. It has been shown that parts of the shelf are permeable to sea water and that the brine soaks in from the seaward edge for a distance of several miles. The problem of how marine debris such as headless fish come to be lying on the surface of the McMurdo Shelf far from the sea has also been answered. It is suggested that the shelf freezes on the bottom while ice ablotes, or evaporates, from the top. Divers have reported that frozzle ice which forms on the sea bottom under the shelf sometimes breaks loose and floats up until it freezes on to the shelf. The frozzle ice contains marine debris which thus gradually works its way to the shelf surface.

For some years a snow pit dug thirty feet into the shelf two miles from Scott Base has been providing data on other aspects. Snow at the bottom of the pit has been dated as falling in 1942 and is uncontaminated by radio-active material from the atomic bombs of World War II. Tiny spring gauges set in the pit walls have been used to measure compaction of snow at various levels. Changes in measure-

25

ments between marker pins in the faces show how their shape has altered during the pit's slow passage towards the sea.

The behaviour of the pressure ridges in front of Scott Base gives more information about the movement of the shelf and the relationship between sea ice and shelf ice. The ridges are formed because the sea ice is jammed between the land and the shelf, which advances seawards at about 300 feet a year. Historically the sea ice in the pressure ridge area has remained fast year by year and has not broken out to float away with the rest of the McMurdo Sound ice. The apparent pattern changed in the late 1960s when at breakout time there was open water in front of Scott Base. But each autumn the ice forms again. Pinched between shelf and land the ice folds and cracks and when the force becomes great enough the edge of the shelf itself is forced to fold as well. In this way the forward movement of the shelf is absorbed until a warm and windy season, which completely frees the Sound of ice, allows the pressure ridge area to float away and the straining process to begin afresh.

The mechanism of folding, movement and absorption of the shelf's flow in the pressure area is still little known and again, the results obtained from a measurement programme will be applied to the understanding of ice behaviour in other parts of Antarctica.

At the other end of the McMurdo Shelf, Cape Crozier, the eastern tip of Ross Island, marks the transition into the Ross Shelf. This is a junction between sea and land where the Ross Shelf is subjected to heavy stress making it a likely place for the calving of icebergs. A small hut withstands the fogs and blizzards for which the Cape is known, and from it New Zealanders and Americans in co-operation continue scientific studies. The site is good for recording the frequency of iceberg calving and studying the shelf's behaviour during the process.

The time taken by an iceberg to disintegrate and melt is another unknown of Antarctic glaciology. Antarctic coasts are littered with standard icebergs which might remain a hundred years before diminishing enough to float again. With bulk a thousand feet below the surface of the sea they drift with deep currents rather than with storms. They may spend seasons trapped in the slow westerly drift of the pack ice, the belt of rafted and broken sea-ice floes which forms a frozen moat about the continent in winter and a loose chain hazardous to ships in summer. Finally free of the continent's influence they drift north towards warmer seas, melting and disintegrating as they go.

The Antarctic seas begin to freeze between late February and April each year. The Ross Sea freezes later than most areas because though it is further south its size is great enough to influence coastal air circulation and therefore climate. Each autumn, new ice grows and old ice, formed the previous autumn and winter, then released from the land in summer to join the pack belt, is consolidated. Sea ice can grow about eight feet thick in one season.

26

Fast ice or bay ice which has remained unbroken because the summer was relatively cold and still, or because it formed undisturbed by heavy gales the previous autumn, grows thicker with another season of freezing and snowfall. The sea freezes outwards or northwards from the land and southwards from the loose pack-ice floes drifting offshore until the cover is complete. Winds, sea currents and pressure from moving icebergs keep the sea ice cracking and moving, folding it in pressure areas and rafting the floes so that they jumble on top of one another. Ships caught in the pack can, like Sir Ernest Shackleton's *Endurance* trapped in the Weddell Sea, be crushed by the pressure of the ice. When spring comes the sea does not refreeze between pressure cracks and the floes remain free to break out and blow or drift away.

Another New Zealand glaciological study has been the breakout of sea ice from McMurdo Sound each spring. Periodic photographic coverage of the state of the Sound's sea ice has revealed a pattern of breakout behaviour which is of great value in planning operations for ships which bring supplies for New Zealanders and Americans each summer. Satellite photographs are now routinely used for prediction of ice conditions.

Before the breakout begins the McMurdo Sound and Cape Hallett sea ice makes an invaluable travelling surface for men, sledges and vehicles. Much of it swept clear of snow, the sea ice surface in spring is a hard milky green, dimpled by the scouring drift. The first aircraft flown from New Zealand by the United States Operation Deep Freeze can safely land on sea ice runways which are firm enough for conventional wheeled aircraft as well as for those equipped with skis. Alternative airstrips are laid on compacted snow on the McMurdo ice shelf and American engineering has even achieved an ice runway on a snow-free shelf area south of McMurdo Sound.

So that cargo ships can enter the Sound early in the summer, the United States Navy employs icebreakers to break a channel south and to escort its own supply ships and H.M.N.Z.S. *Endeavour*. The fast sea ice forms a good berth for shipping, which can moor alongside and discharge cargo onto waiting sledges. Crewmen build fires and eat barbecued steaks on the ice beside their ship, or use the flat surface as a sports field. But modern engineering has made it possible to build Antarctica's first permanent harbour in McMurdo Sound. At Winter Quarters Bay where Captain Scott's first expedition home was built on Hut Point, the United States had built Eliot Quay, a wooden pier where cargo ships can tie up. Wheeled trucks carry their cargo up the volcanic rubble roads of McMurdo Station giving a far more efficient service than tractor-pulled sledge trains so long as the icebreakers can keep open water around the quayside.

In all its forms, ice rules life in Antarctica. Its domination of the surface landscape makes a man curious to explore and investigate its interior just as he does the sea or the air. In a snow cave near Scott Base you can enter the ice and

27

examine its interior substance. The cave is formed by a snowfield cornice which has folded over and down until it has touched the sea ice at the foot of a concave hillside. Here are winding tunnels and crevices, spacious galleries hung with hoar crystals, deep cracks in a clear blue floor of ice through which you imagine you can hear the sea. The ice moves and creaks audibly, threatening collapse. Inside you shiver when the ice cracks and think of turning with your glowing, yellow paraffin light, escaping past the jewelled walls of hoar crystals into daylight. But round about are signs of other explorations in other years. The cave seems safe again and you relax in wonder at the moving ice.

4. *The Land*

DOWN ITS COLD AND STONY BED THE ONYX RIVER BRIEFLY runs in summer, silver-braided in the midnight sun, an oasis of noise in the Antartic silence, a commonplace and careless stream in a land as strange as Xanadu. When the river runs, memories of a warmer, richer life seem to stir in its valley. In melt puddles algae blooms like the Precambrian slime in which life first began on Earth. Bacteria stir through the sparse soils in their blind movement to reassert the dominance of fertility over the land. Lake Vanda's ice sheet melts away to leave a still, blue lake on which the mountains are reflected.

Strong sunlight on the valley rocks casts shadows in ripple marks left by sea waves 350 million years ago and on the wavering tracks of seabed works. Fish fossils are revealed and the skeletons of Devonian shells. Coal seams tell their story of once luxurious vegetation. Penguin bones and pollen grains remain from the time Mt Erebus first erupted in the ocean floor. The Onyx River seems to promise that sometime it could all happen again.

The river's course is through an area unique in Antarctica for it is the largest region that is free of ice. Two thousand five hundred square miles of crumbling brown hills and steep-walled glacial valleys surround the river. They have been released by the receding ice sheet which now thrusts only dripping glacier snouts into the valley heads, leaving the Dry Valleys, a great window on Antarctica's geological past, a door to the discovery of how the continent began.

The three main valleys of this oasis owe their freedom to chance. The conventional explanation is that during a time of volcanic activity 150 million years ago beds of hard rock called dolerite were laid down across their heads. It is suggested that the ice sheet grew to cover these, then shrank until its level was low enough to be largely dammed by the dolerite. The valley glaciers, left unreplenished by the

28

sheet, gradually evaporated. This interpretation is open to argument, but in any case strong summer westerly winds now blow off the sheet into the valleys causing a rise in air pressure at their floors and so a rise in air temperature. In summer the bare rock absorbs solar heat, the small snowfall evaporates and the glacier snouts melt to feed the twenty miles of the Onyx River, together with other, smaller streams which in turn feed fresh water lakes. Chance and the established balance between seasonal heating and cooling has created an ice-free oasis.

New Zealand geologists and geophysicists have visited the Dry Valleys each summer since the beginning of the I.G.Y. following up the work of early investigators with Scott's expeditions. Their work has yielded much essential information about Antarctica's present and historic construction. In the valleys the stratification of rock which records the continent's geological history is clearly revealed, and a control area is provided for survey work done in the rest of the Victoria Land mountain chain and wherever geological formations can be seen free of ice.

Such field investigations are long and arduous, for journeys must be made into Antarctica's most inaccessible places. Where the rocks beckon, the geologists must go, and likewise the geophysicists must strike out over the ice sheet with instruments for discovering the nature of the land beneath the ice. National expeditions must pool their resources with those of other nations because findings from one area must be supplemented by those from another before they can be fully understood.

The geologist enters a colourful and exciting world in Antarctica for in low temperatures rocks and minerals are not subject to the rapid chemical weathering of warmer latitudes and tend to retain the pristine colours of their pure states. Unspoiled by a covering of soil and vegetation, rock strata have a startling clarity of form. The massive folds and faults which gave birth to Antarctica's mountains are seen arrested in hillsides as if they had been struck still by the cold.

Science could not attempt to answer the question of how Antarctica or any other continent was formed unless it had a general theory about the inner nature of the Earth as a whole. It therefore postulates that the Earth has a thin rock crust overlaying a body of less viscous rock called the "mantle". The crustal thickness varies from between eighteen and twenty-five miles for continental land moves to between three and six miles for oceanic areas. Within this shell the Earth's rock is pulled this way and that by the moon's gravitation. Because of the moon, tides like those of the ocean are created in the Earth itself.

Where earthquakes occur they are generally within the crust but in some cases they originate as deep as 450 miles down, their shock waves pulsing to the surface, and out through the crust. Because it "floats" on the mantle, the crust can be contorted, sinking into the mantle in response to loading by ice or the deposition of sediments by moving water, rivers or oceans. At times, molten rock trapped in the crust forces its way to the surface to create a volcano like Mt Erebus and its extinct or dormant sisters on the Victoria Land coast, or to force its way between

29

rock strata under the surface like the dolerite of the Dry Valleys. At other times expansion and contraction in the crust create folds which are thrust up to form mountains and down to form deep valleys; or whole blocks of crust, split along fault lines, are raised bodily above the surrounding crust, or sink to form broad, square-walled valleys.

Gravity measurements are a useful tool for investigating the nature of the crust. Gravity is the attraction which all masses have for one another (most familiarly manifested by falling bodies) and measurements of its force at various points on the Earth's surface are based on deviation from a "normal" value. Thus a gravity value less than the normal is recorded at a point where the Earth's crust is overlain by a thick ice load because the density of the ice under the gravity meter is less than the density of crustal rock. Gravity measurements taken over areas of dense, or heavy rock are high; and if the nature of the rock base is obscured by ice, measurements of the strength of the Earth's magnetic field must also be taken at the survey station, for they give an indication of the type of rock to which the gravity meter is responding.

Into this known background of the Earth's behaviour Antarctica must fit but exactly how it fits is still being discovered. It is even quite apparent that the completed record of Antarctica's tectonic, or construction history may overturn some long-held theories of the world's past and explain many problems of the past in other continents.

Because of the ice sheet it is likely that Antarctica will always keep some secrets of its birth. About ninety-eight per cent of the land is ice-covered. Deep drilling can record the nature of the rock under the surface in ice-free latitudes but is impossible where an ice sheet covers the land.

Five techniques remain available for investigating the land beneath the ice. Geologists examine rock outcrops such as those in the Dry Valleys. Seismologists explode dynamite in the ice sheet and record the time which the resulting shock waves take to travel between the surface and the rock beneath by which they are reflected. This is called reflection seismic shooting, while in refraction shooting the passage of the shock waves through the rock between two recording stations on the ice surface above is recorded. The velocity with which the refracted waves travel indicates the nature of the rock through which they are passing.

Measurements of the force of gravity at various points on the ice sheet, the elevation of which is known, indicate ice thickness and thus the height of the land beneath. Records of the way in which seismic or earthquake waves originating in other parts of the world travel through the Antarctic crust tell its thickness. Radar can be used to penetrate the ice to bedrock level.

Before the I.G.Y., little was known about the shape of Antarctica under the ice, but the long vehicle and dog sledge traverses which have criss-crossed the Polar Plateau and mountain areas since then, doing seismic depth sounding, gravity and magnetic surveys, and geological investigations, have established the broad outlines of the land. Earthquake recording at permanent bases has supplemented this knowledge.

The shape of Antarctica is now emerging and is seen to consist of the broad continental shield area of East Antarctica with a rock crust of thickness typical of other continents, and a series of islands in West Antarctica of somewhat thinner crustal thickness standing out of a crust area of oceanic thickness. On the division between the plateau-like East and the ice-inundated islands of the West occurs the Trans-Antarctic Range, a 1,400 mile line of folded and block-faulted mountains stretching from Cape Adare, the northern tip of Victoria Land, south to within 300 miles of the Pole, and apparently continuing on again northwards to the southwest of the Weddell Sea. From the Atlantic side of the continent, the South American Andes dip under the sea as the Georgia Arc and emerge as the Antarctic Peninsula, Antarctica's furthest north. The fate of the "Antarctandes" is not yet known but the range appears related to a fan-like formation thrusting deep into West Antarctica.

The Trans-Antarctic Range has all the glory of the world's great alpine areas. Though the peaks curve away slowly on their landward side, they tower 11,000 and 12,000 feet over the sea coasts of Victoria Land and climb even higher at their southern end. It was once thought that a submarine trough must connect the deep embayments of the Ross and Weddell Seas but most recent investigation indicates that this is not so. An ice-filled trench, the result of the mountain building process, appears to lie along much of the division between East and West but does not connect the two seas.

About the shore of Antarctica, a continental shelf extends into the ocean and is, curiously, at a much greater depth than that around other continents. It was once thought that this was because the Antarctica land was deeply depressed by the ice sheet's weight but gravity studies have shown that while the interior is depressed, the coastal and offshore regions remain at a normal level.

If the ice sheet were removed, the land would not immediately spring back into place. The process would be slow, measured in inches a year. The end result would be a rise of about 2,000 feet, bringing most of East Antarctica above sea level except for a low lying basin in North Victoria Land. The Plateau land area would be almost a million square miles, from a few hundred to a thousand feet high, sloping gently from west to east. The major part of West Antarctica would remain below the sea but some large islands would project, along with an undetermined amount of land under the remote and still unexplored areas of the West Antarctic ice sheet.

As in the other continents, Antarctica's earliest rocks are from the Precambrian age, the great era of continental building which geology dates as beginning about 3,200 million years ago and lasting about 2,600 million years. Along the Trans-Antarctic Range the rock strata laid down in successive ages of continental

31

building are sometimes bared. Their early sequence has unexpectedly been found to have much in common with the stratification and thus the geological history of other continents. The rock type dates the strata of the early ages, and later formations bearing fossil plants and marine animals can also be dated according to the type of fossil found. Where no fossils exist, or where their age must be checked, samples can sometimes be dated by measurement of the concentration of radioactive isotopes of the chemicals they contain. Here a newly developed technique is proving useful, and is given impetus towards refinement by the demanding nature of Antarctic problems. Work such as this must be done in the laboratories of low-latitude countries and the international co-operation of science allows specialists of one nation to assist in the analysis of samples from another nation's expedition.

Because of Antarctica's ice cover, the geological history of the Precambrian era of its building still holds some major unknowns particularly in the tectonic relationship between early, or basement rocks in one ice-free area and those in an area obscured by ice. But the general pattern of Antarctica's building is emerging.

In the Precambrian age there were many periods of mountain building through folding, the upthrust of molten rock through the crust as volcanoes and lava flows and the deposition of sediments from water erosion. The Cambrian age continued with the deposition of marine sediments for 100 million years. For a similar period granite rocks, which cooled slowly in large crystals, intruded into the crust—the glowing pink granites now common on the western shore of McMurdo Sound. Erosion followed and the land was again uplifted. Erosion and deposition of sediments continued through the Devonian and Carboniferous ages until about 300 million years ago. Thus the famous Beacon sandstone beds of Victoria Land were laid down containing their plant fossils, shells and even fish.

Antarctica's first recorded glaciation came next laying down a strata of tillite, or glacially deposited conglomerate rock. Fifty million years later Antarctica was flourishing and warm. Animals left tracks and skeletons in the mud of a wide sea-fringe plain of dunes, lakes and estuaries. The common plant, *Glossopteris*, with its long and fine-veined leaf grew thickly and was later turned to coal. Rivers ran and flooded, laying down silt which became another layer of Beacon sandstone. Between 200 and 150 million years ago there was volcanic activity and the dolerites so striking at the Dry Valleys forced their way between the sandstone strata.

About 100 million years ago it was again warm and a variety of shellfish and plants flourished about the Antarctic Peninsula building the land as they died. Fifty million years ago the volcanoes of Victoria Land began to emerge. This was a great time of convulsion when Mt Erebus was born, the Andes grew south under the sea into Antarctica, the crust cracked and folded heaving up the Trans-Antarctic Range. Bird bones were buried, and pollen from plant flowers. Ten million years ago the ice sheet grew.

One of the immediate problems of geology in Victoria Land is to distinguish between two distinct units of basement rock—an upper limestone strata mildly metamorphosed or transformed by heat and pressure and at least in part of Cambrian age; and a lower, older bed of similar though more metamorphosed rocks. Establishing the gradation between them in various areas is currently one of the chief tasks of New Zealand's Antarctic geologists.

In the Antarctic mountains many of the geological formations are easily distinguishable. Flat-topped or sharp-peaked nunataks or rock outcrops stick up in the ice sheet bearing the marks of glaciation from an age when the sheet was many thousands of feet thicker. Their sheer sides reveal rock strata in many-coloured bands, the pale gold of beacon sandstone sandwiching black dolerite or overlain by red lava flows. A split rock may reveal the fossil leaf of the *Glossopteris*, pollen grains or shell. A boulder-littered hillside may bear a forest of fossil tree stumps. In North Victoria Land the nunataks can be seen as ice-eroded island remnants of the great Kukri Peneplain, once the deposition bed of the Paleozoic age Antarctic river sediments.

The influence of wind and cold on Antarctica's land can best be studied in the McMurdo Oasis and similar areas. Here, where a post-glacial era has begun, soils are beginning to form and summer travellers trudge across the duney slopes with dust clouds round their feet. Lichens and mosses grow on the rocks, and algae, a simple slime-like vegetable, grows in the lakes and melt pools. The lakes themselves, though covered with twelve or more feet of ice have paradoxically warm water near their floors. At first this was thought to be caused by geothermal heating. Later studies have shown that the warm bottom waters are very salty and that well-defined layers of temperature and salinity are maintained. It is thought that solar radiation contributes most of the heat stored in the Dry Valley lakes. The salt-tense water layers inhibit convection currents which might dissipate the heat through the whole lake contents and so it has accumulated over the years in the most dense bottom water.

Wind-blown sand has worn the rocks into curious shapes called ventifacts, curves and hollows, leaf-like projections and mushroom bulbs of pleasing proportion. Constant frost has created a fishnet pattern of cracks on the rubbly ground called frost polygones or patterned ground. Its formation is not fully understood but is attributable to the continued expansion and contraction of the surface under varying temperatures.

There is still plenty of room in geology for a wholly satisfactory theory explaining why Antarctica was once warm. It need not always have been a polar continent although there is no reason why the plants found fossilised could not have existed well-adapted to the six months dark and six months light of a polar year at a time of world-wide climatic warmth. There is certain evidence that the geographical and magnetic poles were not always in their present position. Paleo-

33

magnetic studies or studies of magnetism left in old rocks show by the direction of its lines of force that the rock crystals must have formed when the Earth's lines to magnetic force converged on a Pole in a very different area. The South Magnetic Pole may once have resided in the region of Tonga Island in the South Pacific.

Another possibility is that postulated by the theory of continental drift, an explanation out of favour until the paleomagnetic studies of the early 1950's forced its reconsideration. Antarctic geological discoveries since the I.G.Y. have given impetus to its reappraisal. The continental drift theory states that once Antarctica was part of a super-continent, called Gondwanaland, in which it joined with Australia, India, Africa and South America. It says that somehow Gondwanaland split up into the now separate continents which, because of their great crustal thickness, floated apart on the plastic rock of the Earth's mantle until they took up their present positions. The geological record revealed by Antarctica's rock strata with its basement rocks, fossil bearing sandstones and glacially deposited tillite, tallies remarkably with the stratigraphy of the other Gondwana continents.

Thus it is proposed that at the time when Antarctica was warm and fruitful it was part of a much larger land mass where the same processes continued under the same climatic conditions until Antarctica wandered to a colder home. At present, science has agreed not to dogmatise on these ideas, to continue the patient exploration until sufficient is known to settle the question. The propositions of world-wide warmth and polar wandering suffice to account for Antarctica's fossil record but they cannot account for its similarities with the other Gondwana continents. Furthermore there is evidence that the continents are still moving at a rate of about 3 cm a year.

The New Zealand Department of Scientific and Industrial Research Geological Survey and Geophysics Division have, together with the Victoria University of Wellington Antarctic Expeditions, been the agencies for New Zealanders' investigation of Antarctica's land both in the field and at Scott Base. The University has sent field expeditions to the Dry Valleys and other areas each summer from the beginning of the I.G.Y. Geologists accompany all field survey parties which explore the Trans-Antarctic Range in Victoria Land and now the process of initial exploration is complete are engaged in special investigatory projects in local areas known to harbour geological information on particular problems. Their contribution to Antarctic geology is unequalled.

So far, neither they nor the geologists of any other nation have found minerals of sufficient concentration and commercial value to be worthwhile exploiting. Coal seams of commercial rank and quality have been found in the Trans-Antarctic Range, but coal is not precious enough to make its extremely difficult commercial extraction worthwhile. This consideration would prevent any mineral resources except those of great value and scarcity from being used in Antarctica. There is still the possibility that such a resource may be found but the slight chance provides little additional impetus for geological research in Antarctica.

New Zealand's gravity survey work done by the Geophysics Division has been entirely of a reconnaissance nature to give a network of values for the force of gravity at various points in Victoria Land. The division is also undertaking both station and field seismological observations. For some reason no earthquakes of significant dimension have been recorded as occurring in the Antarctic area and the reason for this remains a geophysical problem of the first order. Perhaps it is so because the weight of the ice sheet inhibits movement in the Earth's crust.

Both at Scott Base and Vanda Station the Division runs seismometers to keep a continuous record of seismic disturbances both in Antarctica, if any, and coming in from all parts of the Earth. The lack of earthquakes at Scott Base is found all the more surprising because it is in the middle of an area of current volcanic activity. The recording instruments used are provided by the United States Coast and Geodetic Survey and are part of a world-wide network of similar standard instruments in many countries. Information is recorded by seismographs which print the tremor as a wave-line on a chart. It can then be sent to similar stations all over the world to assist research.

Information from Victoria Land is particularly important, for the New Zealand stations are the only ones on the southern edge of the great circle of seismic activity which encompasses the whole Pacific Ocean and runs right through New Zealand itself. Just as important is the fact that these stations lie on the dividing line between East and West Antarctica and can thus scan either side of the continent from a central point. Antarctic records provide crucial information on the earthquakes' points of origin and the speeds at which their waves have travelled through the Earth's crust. It is this type of recording which has enabled the Division's scientists to estimate the thickness of the crust in Antarctica.

The Scott Base seismographs also commonly record small and apparently local vibrations in the earth with a frequency which seems to be peculiar to the area. These vibrations have been interpreted as being set up by icebergs breaking off the edge of the Ross Ice Shelf near Ross Island. Given the name of *icequakes* they are being investigated by a specially developed type of seismometer which can be set up at field stations to record the vibrations as noise on a magnetic tape.

The seismographs also record micro-seisms created in the Earth's crust by pressure differences resulting from the interaction of earth, air and water. They are accentuated to some degree by waves beating on the Antarctic shores with a force increased by the cyclones which roll round the continent at great speed, undeflected in the landless Southern Ocean.

Problems surround the scientist on every hand in Antarctica and lie beneath the ground on which the Scott Base huts rest. Temperature measurements in the earth down to fifteen feet show that the geothermal gradient, or increase in temperature with depth, is extraordinarily high—in fact the rate of change is forty times greater than normal in the surrounding ground. This is probably due to the closeness of the volcanic activity of Mt Erebus. A current New Zealand project is to drill down to at least fifty feet to find how the temperature gradient continues to vary and perhaps what causes the variation.

Antarctica's earth is still alive and rumbling, moving with infinite slowness through the next great changes of its growth or collapse, creation or decay. The hot earth under Scott Base may indicate the last influence of dying volcanism or the beginning of a great new cycle which, in a hundred million years, may cause the continent to take another shape or the whole world's moving skin to change.

5. *The Day*

AT MIDSUMMER MIDNIGHT THE WORLD IS STILL. WITH A GOLDEN glow of sunlight on his arms and face the Scott Base nightwatchman observes the weather, pauses to listen to the domestic beat of generator motors and continues his careful rounds. At the dog lines skua gulls hover and mew over their scraps of seal blubber. The dogs whine in sleep or stretch and yawn, their iced chains clinking. At McMurdo Station the rubbish cart rumbles behind a bulldozer through dusty streets. At Cape Royds, two thousand penguins doze and blink while skuas crow savagely to the Sun with outstretched wings. Seals snore and wheeze in the shallow beds hollowed by their warmth beside the tide cracks of the Sound. High among the mountains of the Trans-Antarctic Range travellers sleep enclosed in thick, down sleeping bags, their breath freezing in inch-long crystals about their faces.

At midnight the world is blue and gold with long shadows in the pressure ridges. The huge western glaciers are molten flows of orange light. Five miles across the ice from Scott Base the aircraft and buildings of Williams Field are raised up by the mirage into a city of skyscrapers and spacious buildings, parks and boulevards. The snow-fields on the dome of Erebus are darkly shadowed, and the volcanic smoke plume streams down a summit wind. The Sun turns the tall aerial masts which crowd the slopes behind Scott Base into golden spears irradiated with the solar power which they seek to describe. At midnight the Sun asserts its government of the Earth with light.

Inside the base the instruments which investigate its influence run on. With radio beams pulsed through their aerials they record the impact of solar energy on the thin ionospheric gas thirty miles high, and higher. They listen for radio noise generated by thunderstorms in warm latitudes. They record the continuous variation of the Earth's magnetic field and the ebb and flow of electric currents in

36

the Earth itself. At Second Crater five miles away space satellite signal reception aerials probe the frosty air waiting for the next orbit of a Beacon satellite with its messages from space about the state of the ionosphere.

These are the instruments of geophysical investigation in the upper atmosphere. They seek to define the relationship of our planet to the Sun, the nature of a solar atmosphere as broad as our Constellation, the nature of the unseen forces in man's immediate physical environment. With similar instruments at Vanda Station they are part of a worldwide network of geophysical research stations observing the same phenomena at the same time and according to the same criteria, that is synoptically.

Along with most other branches of Antarctic science, geophysical studies of the upper atmosphere were begun by New Zealand during the I.G.Y. and still reflect the international nature of such studies. Historically, they belong at the head of Antarctic scientific inquiry. Although upper atmosphere phenomena were not then linked with terrestrial magnetism Sir James Clark Ross, who had discovered the North Magnetic Pole, made the first voyage south beyond the confines of the pack ice in an attempt to locate its southern counterpart and study related phenomena. His voyage of discovery was as firmly linked with science as the field journeys made today, even though his primary purpose was thwarted because the Magnetic Pole then resided far inland.

The great leap forward of twentieth century science demanded more data about the upper atmosphere and the Sun than centuries of visual observation and mechanical experiment had been able to furnish. The advent of the science of electronics made accurate measurement and automatic recording of geophysical phenomena possible for the first time.

Antarctica fits into the global network of geophysical studies with special pertinence. It offers a very clear atmosphere for visual study such as that of the aurora and a stable platform, unlike the shifting sea ice of the North Polar area, on which to erect instruments; a population sparse enough to be almost disregarded in the planning of rocket flights for upper atmosphere research; and little of the background radio noise which can, in populated areas, affect clarity in recordings of radio wave phenomena. Above all the Antarctic year is split into roughly six months day and six months night; this greatly increases the range of the Sun's influence on the polar atmosphere and, in the region of the South Magnetic Pole, much upper atmosphere activity is concentrated by the Earth's converging magnetic field.

Just as in its study of the nature of the Earth itself, geophysics must have a general theory of the Sun and its effect on and interaction with the Earth. Earth satellites and space probes have made the first satisfactory projections of such a theory possible, and space science in which measurements and observations are made easier by the accessibility of the subject Earth. The I.G.Y. set in train a series of fundamental discoveries about the planet Earth's environment in space. Space research has continued where land-based projects left off and since the I.G.Y. a full eleven-year sunspot cycle has been observed. The result has been a revolution in man's approach to the problems of understanding solar processes and the nature of solar energy.

Science proposes then that there is a Sun whose atmosphere of gaseous particles extends, tenuous and sparse yet distinct, far beyond the farthest planet. The Earth spins as it moves round the Sun through the Sun's atmosphere. The Earth also has an atmosphere ranging in density from the air we breathe at its surface to the thin heights of the high upper atmosphere at least 40,000 miles distant in the sunward direction. Gases in this region are so far removed from the Earth that they do not spin with it, although they still move bodily with it round the Sun.

The Sun presents to view a core of burning hydrogen called its disc. The edge of the disc is called the corona. As it burns, the sun steadily ejects a "wind" of gas which rushes at supersonic speed towards the planets, eveloping them in a constant shower of energy. At times sunspots or solar flares create bursts of energy like a thunderstorm. The frequency of solar storms varies according to an eleven-year cycle between maximum and minimum activity but the solar wind blows continuously. The I.G.Y. marked the period of maximum sunspot activity, the International Quiet Sun Year the minimum period eleven years later.

Like the Earth, the Sun has an electro-magnetic field which fluctuates in intensity and causes fluctuations in the Earth's field. The Earth's field is generated by electric currents moving deep within its core. The result is a doughnut shape of lines of force radiating out from both the Earth's magnetic poles which constitute the entrances to the hole in the doughnut. Clouds of gas, called solar plasmas, emitted by the Sun also transport their own magnetic fields bodily through space, and the relationships between the magnetic fields of Earth, Sun and solar plasmas is gradually being defined as space resarch progresses.

The Earth's gravity field keeps the tenuous upper atmosphere in position while the magnetic field helps to trap the atomic and molecular gas particles emitted by the Sun. It is thought that the particles spiral up and down along the Earth's lines of magnetic force from pole to pole reversing their direction at either end when the lines of magnetic force become sufficiently concentrated. They form the Van Allen radiation belts, those spectacular space discoveries of the I.G.Y.

The points on the Earth's surface at either end of one line of force are called conjugate points, and there appears to be a relationship between the flow of electric currents in the Earth's surface near opposing conjugate points. The lines of force also affect cosmic rays—atomic particles travelling through space and estimated to contain, in their sum, half the total energy of the Universe—by bending their paths towards the Earth where their incidence can be recorded. Cosmic ray incidence is also associated with solar activity.

The Sun also emits energy in the form of electro-magnetic waves. The Sun's disc emits ultra-violet rays and the Sun's corona emits X-rays and gamma rays. The ultra-violet rays penetrate to the Earth's surface but the gamma rays are absorbed by the upper atmosphere. There they ionise, or excite, the gas atoms which gain or lose an electron. Such ionised gases form the ionosphere which itself is composed of different ionised layers known as the D,E and F layers or regions according to their degree of ionisation.

In this excited state some of the ionised particles emit energy in the form of light seen spectacularly as the aurora—most common in the polar regions because excited particles tend to concentrate where the solar wind influence is strongest.

In short, the Earth in space is somewhat like the soft iron bar inside the coiled wire of the familiar schoolroom physics experiment. When an electric current is passed through the coil the bar becomes a magnet. If the bar is magnetised and then put inside the coil a current flows in the wire. Such is the behaviour of the Earth inside the doughnut cell of ionosphere and magnetic field. The aurora may be likened to a neon light tube in which gas particles emit light when an electric current is switched on and passed through the tube.

Space research since the I.G.Y. has shown however that the doughnut model is appropriate only to the region relatively close to the Earth. Outside this region which is tightly controlled by Earth's magnetic field, the outer atmosphere appears to be shaped like a gigantic teardrop, its blunt end facing the Sun and its tail streaming out towards the further planets, as the solar wind rushes by. The leading edge of the "teardrop" is perhaps 40,000 miles from Earth on the sunward side and the end of the tail perhaps 4,000,000 miles away in space. The lines of magnetic force are bent in the same shape as the atmospheric gas. It is thought that the tail of the teardrop "flops" in the solar wind, allowing sun-charged particles to enter the Earth system and spiral their way into the ionosphere. Because the lines of force with conjugate points at the poles are those most bent askew by the solar wind, the incoming solar particle energy concentrates near the poles creating the auroral region. Direct radiation naturally penetrates the Earth's sunward side atmosphere without being affected by the atmospheric distortion caused by the solar wind.

The ionospheric layers close to the Earth are of prime importance to man's everyday life for they affect the way in which radio waves travel. Though the ionosphere is curved like a spherical roof over the whole Earth its underside is irregular. Radio communication waves travel by bouncing back and forth between the ionised layers and the Earth by ricocheting across the uneven under-surface of the layer itself. The distance over which a wave of a particular frequency can be transmitted is governed by the uniformity of ionization of the layer along which it takes its bouncing path. If there is a "hole" in the ionosphere or if one of the lower layers is totally absent because radiation from the Sun has not ionised the gas at that level to form a reflecting surface, a radio signal transmitted at, say, Scott Base, may escape out of the Earth's atmosphere and dissipate into space.

Because the South Polar ionosphere is sunless in winter, many "holes" develop in the ionosphere and the Scott Base or Vanda Station staff often find it impossible to have their weekly radio telephone talks with families in New Zealand.

New Zealand's methods of observing the behaviour of the upper atmosphere in the Antarctic are usually restricted to ground-based electronic instruments but other nations, notably the United States, are continuing balloon, rocket and satellite observations at different atmospheric levels. New Zealand scientists collaborate in these projects. The Beacon satellite receiving equipment at Second Crater is run for the University of Auckland and is one of a line of similar stations stretching south from the Cook Islands through New Zealand and the sub-Antarctic. Fourteen times a day the satellite passes overhead, its signals providing information about the state of the ionospheric particle clouds as they are distorted on their Earthward path. The problems the satellite is helping to solve are many. There is evidence that mighty "winds" blow in the ionosphere, moving great volumes of ionised particles about with astonishing speed; but their nature is barely known. The ionospheric wind can be observed affecting the trails of meteorites, but this is a chancy method now being replaced by observation of rockets which emit smoke clouds, the movement of which is plotted. There is evidence that sheet-like electrical currents are induced and dissipate in the upper atmosphere, moving freely and with great speed: but the mechanics of their cause and behaviour is still little known.

Apart from their special value in the observation of these phenomena, Antarctic geophysical research stations make possible a planetary view of the environmental forces which surround us and which we need to understand in the search for the origins of life and energy.

The Sun's magnetic field has been observed to undergo a complete reversal of polarity—its North Pole became its South Pole. It is possible that the same thing could have happened to the Earth at some time in the past, exposing our atmosphere to cosmic particle and radiation bombardment while the protecting magnetic field was absent. The effect of such an experience on biological life is, at the moment, incalculable. Basic scientific research in the upper atmosphere is showing man that environment is wider than the air he breathes and the ground on which he walks, and New Zealand's Antarctic stations are contributing their share of information.

Scott Base and Vanda Station have a further advantage for upper atmosphere research for their longitude is very close to that of New Zealand's. The result is a chain of observatories stretching roughly in a line from the New Zealand station at Rarotonga, in the Pacific Islands, through three stations in New Zealand South Island, Campbell Island in the Sub-Antarctic, McMurdo Sound and ending with the United States observatory installations at Amundsen-Scott Station on the South Pole.

Nowhere else south of the Equator is there a chain of such stations covering almost ninety degrees of latitude lying on roughly the same meridian of longitude and thus keeping the same time.

The polar ionosphere gains from the course of the Sun, the characteristics which distinguish it from the low latitude ionosphere. The Arctic and Antarctic circles mark the latitude at which the Sun does not set at mid-summer and does not rise at mid-winter. This is the phenomenon which first gave the polar regions their names. The Greeks called the northern lands Artikos because there the constellation of the Bear whirled round the horizon at night and did not set. The opposite end of the Earth became the *Antarctic* in the English language. In late winter the Sun first rises above the horizon in the latitude of Scott Base about 20 August and remains above the horizon continually from about 27 October to 15 February, setting again for the last time about 23 April.

The permanence of summer sunshine means that the polar ionosphere maintains a higher level of ionisation in summer than that of lower latitudes which is subject to a diurnal variation because it is shaded from the Sun's rays at night. Conversely the polar ionosphere undergoes much greater changes during the long polar night than occur in the short low-latitude night. It has been found however that, rather than disappearing altogether in the absence of sunlight, the polar layers of ionisation break up into cloud formations which persist throughout the winter by a mechanism as yet unexplained.

Instruments for upper atmosphere research at Scott Base and Vanda Station are operated for the Geophysics Division and the Vanda Station Laboratory of the D.S.I.R. They give a continual record of ionospheric behaviour which is of particular interest following solar events. The effects of an event such as a solar flare can be observed within hours of its occurrence because the rays and gases emitted during the flare travel with great speed. As well as pouring energy directly into the ionosphere a flare can cause violent fluctuations known as magnetic storms in the Earth's magnetic field, either because of attendant fluctuations in the Sun's magnetism, or because of the passing electro-magnetic field associated with the ejected cloud of particles.

Chief among the upper atmosphere research instruments at Scott Base is the panoramic ionosonde—a device like a marine echo sounder which periodically beams radio signals upwards from the lofty aerials which surround the base. The signals are transmitted for a short period every five minutes over a rising wave frequency. They continue to echo back to the instrument, bouncing off the ionised layer, until a frequency high enough to pass through the ionized layer without echoing back is reached. The echoes are reported on a screen like a radar screen and their image, scaled against altitude and wave frequency, is photographed by

41

a cine-camera. Apart from their value to basic research the records contribute to a world-wide picture of the state of the ionosphere which can be used to predict the most successful radio communications frequency for use in particular regions at various times of the year.

Another aspect of upper atmosphere observation is the recording of "whistlers". When a thunderstorm occurs part of the electrical energy from lightning flashes is released in the radio frequency wave spectrum. In other words the lightning flash makes a noise audible to a very low frequency radio receiver. At Scott Base in 1958 the first very high latitude recording of a whistler was made, a discovery which had important implications for upper atmosphere physics. It is now thought that when energy from the lightning flash escapes into the upper atmosphere the resultant whistler travels along a line of magnetic force to give a descending whistle on the radio receiver.

The first interpretation of the 1958 whistler was that since the lines of force which reach the Earth at Scott Base and its conjugate point at Shepherd Bay, Alaska, must curve out into space a distance of many earth radii it followed that matter must exist, however tenuous, at the point furthest from the Earth otherwise the whistler transmission could not have taken place. It had previously been thought that whistlers could not be received at high latitudes because there were no atmospheric gas particles far out in space to support their passage. High latitude whistlers were accepted as a new tool for the exploration of the atmosphere remote from Earth. Space research later showed, however, that the mechanism was not so simple. It is now known that the high-latitude whistlers are caused by mid-latitude thunderstorms. The radio energy travels along a line of force to its conjugate point in the alternate hemisphere where it looks around the ionosphere to be heard on a polar region receiver. A "whistler" recorder in the Scott Base laboratory listens to very low frequency atmospheric noise for the D.S.I.R. Physics and Engineering Laboratory. At Vanda Station Otago University has installed a receiver to listen to atmospheric "hiss" in the U.L.F. radio range.

Also in these laboratories are instruments recording fluctuations in geomagnetism for the Magnetic Survey of D.S.I.R. Their aim is to provide data to help establish the pattern of the Earth's magnetic variations. It is known that there is a recurrent fluctuation in the Earth's magnetism over the diurnal period, a twenty-seven day period, a sunspot cycle and even a long term or secular change as yet undefined. The pattern of Antarctic magnetic changes is much the same as that of other regions but inclined to be more intense. On a given day magnetic disturbance at the South Pole in summer is greater than the same disturbance observed at the North Pole which is in winter dark.

Geomagnetic studies are also related to events beneath the surface of the Earth—the fluctuations of electric currents in the Earth generated by events in the ionosphere. Scott Base has instruments which measure the difference in electrical potential between electrodes buried in the earth or ice and thus the rate of current flow between them. Data from this equipment can be related to that on earth current changes at the conjugate point in Shepherd Bay, Alaska, and is made available for study by the Geophysics Institute of Alaska.

The Magnetic Survey also engages in field magnetic observations concentrating on the oceanic area between New Zealand and McMurdo Sound. A magnetometer towed behind a ship indicates changes in magnetic field strength along her course. The Survey has also carried out the most recent observation of the position of the South Magnetic Pole or dip pole (the point at which the lines of force bend vertically into the Earth) which alters its position from year to year and even diurnally on an elliptical path that is longer on magnetically disturbed days. The South Magnetic Pole does not lie opposite to the North Magnetic Pole and was last located by a New Zealand party in Commonwealth Bay, 1,000 miles from the South Geographic Pole in latitude 67°03'S and 140°00'E. By now the magnetic pole has wandered many miles from that position which was observed from the top of an iceberg stranded in the bay in February 1962. The new Zealanders' visit was the most recent in a series by scientists of various nations made since Douglas Mawson built the base camp for his Australian Antarctic Expedition of 1911–14 on the shores of the bay.

The men who use the tools of geophysical research at Scott Base and Vanda Station must stay at their appointed stations for a whole year. During the summer when transport moves between New Zealand and McMurdo Sound they can send some data back to the research scientists working in their laboratories in New Zealand. Some of the records like the ionosonde films must be analysed at the base and the data coded and transmitted to New Zealand by radio. It is then sent on to the World Data Centres set up initially in various countries as storehouses for the information collected during the I.G.Y. and still being used.

For the technicians who run the base laboratories this means many hours of looking at ionosonde cine film to note the data in symbols standardised throughout the world. They must also maintain the complex electronic devices which do the recording, working always under the pressure caused by the need to keep their instruments synchronised with similar ones all over the world. Their patient summer world is governed by the Sun, and when the night comes they will turn to its other aspect, the darkness.

Just as its unseen forces, its radiations and electric influence, control the planet's physical environment, the polar sunlight imparts a different rhythm to life for men and animals. With no marked divisions in the day, life can proceed with an apparent freedom untrammelled by the dark. It seems less natural to sleep at "night" and the Antarctic worker is inclined to sleep when he is tired and work when he is refreshed rather than observe a division which seems artificial. At times the Sun requires long periods of wakefulness for outdoor work such as the unloading of a ship. The erection of another piece of scientific equipment demands completion in as short a time as possible rather than according to the dictates of an industrial forty-hour week. The weather may change to obliterate hours of labour with a blizzard, the ice may float away bearing a precious cargo. Outdoor work maintenance of buildings and preparation of winter stores must be concentrated in summer for little can be done in the winter darkness. Time loses meaning in the face of hasty preparation.

The summer is a tumultuous affair, crowded with men setting off on and returning from their journeys and research projects, urgent with the need to carry out their plans while the sun lasts and Antarctica's barriers are down. The men who are to stay behind long for peace and the night for they know that they have yet to achieve their deepest experience of Antarctica.

6. *The Night*

THE FREEZING OF THE SEA IS LIKE THE CLOSING OF EYELIDS IN sleep, a relaxed and languid calm in which the water seems to take on flesh, a rippling surface of skin, golden and glowing in the low autumn sun. The long-rolling limbs of the sea are still and the passage of wind becomes soundless. Homeward bound before winter, ships move through the freezing waters with a rustling sound like the rubbing of old, dry cloth.

McMurdo Sound begins to freeze late in February. In the black water you can see the ice begin from the iridescence of the first minute crystals which slowly grow in number until the surface is opaque. Long streaks of ice plasma are drawn out by the wind over unfrozen water until the whole surface is obedient to the frost.

Then the granular and fluid skin of ice coagulates in small discs like waterlily pads which jostle one another until their edges curl. Rafted together by the wind the pads form larger floes. For weeks in the unstable autumn climate the wind creates an ebb and flow of new ice in the Sound. At times of open water sea smoke forms like steam, whirling across the surface as the warmer air condenses over the colder water. But as the Sun prepares to spin west below Mt Discovery and set for the last time the cracks between the new floes freeze until the sea is flat and solid. The ice begins its steady winter growth to eight or ten foot thickness while its surface is wind-polished to a hard, milky green.

In the dividing weeks between day and night men make their last hurried efforts to complete the tasks of summer. Towards mid-March the last ship must leave or be trapped in the freezing sea. The last farewells are said at dusk.

Suddenly, Scott Base becomes a quiet place. The dozen men left behind are to spend six months alone with occasional contact with their American neighbours "over the hill". For the scientists the routines will continue with the added work of operating and maintaining the instruments which observe the aurora. For the surveyors, geologists, and field assistants the year's action is finished. They will still go sledging but will be confined to short runs to exercise the dogs.

Before the permanent darkness falls they must complete cutting up and stockpiling the summer's kill of seals for winter dog food. The frozen carcasses are quartered and cut in blocks with a power saw. Later in the night sledges have to be relashed, new dog harness made or old harness repaired. All the equipment of sledging and ice travel tents, sleeping bags, boots, crampons, primus stoves, pressure cookers, skis, air beds, ice axes, ropes, mileage meters, fuel containers—must be inspected and repaired in preparation for next summer's field work. Ration boxes must be packed with their carefully weighed food units. Sketch maps of areas to be explored in the coming season must be drawn from aerial reconnaissance photographs taken the previous summer.

Before the night comes the base technical staff—carpenter, engineer, electrician, cook and radio operator—check all the equipment of their departments, making sure it is accessible in bad weather. The carpenter battens down moveable objects against winter blizzards, makes sure that the steel wire tie-downs which hold the huts against the wind are stressed. Fire extinguishers and alarm systems are checked and re-checked for winter is the time of greatest fire danger.

One of the autumn tasks is base refuelling—the storage of petrol and diesel fuel to service the winter needs of vehicles and the base generator motors which supply electric power. Fuel is purchased from Operation Deep Freeze supplies.

New Zealand's winter party in Antarctica normally consists of five scientists or scientific technicians at Vanda and a staff of twelve at Scott Base including the Leader, scientific leader and staff of three, biologist, maintenance engineer, electrician, postmaster, radio technician, carpenter, cook, surveyor, field assistant, dog handler, and storeman.

Only essential outdoor work is done during the winter. Early in the darkness come the year's worst gales, sometimes of eighty knots or more. Blizzard snow drifts pile up on the leeward side of the base buildings until they stretch level from the flat roofs to the slopes of Crater Hill. Temperatures drop steadily reaching their lowest of more than minus 70°F during late July. At this time a cupful of boiling water flung into the air freezes instantly with a loud crackle. The winter world is clean, for the snow buries all the old scars of summer until the huts with their tiny lighted windows and outdoor lights, glowing from kerosene tin shades over snow melter hatches and exit doors, look like green and yellow candy houses surrounded by white icing.

And the winter is a time of danger. In darkness and swirling drift snow a man can be lost a few yards from the safety of buildings. A short journey to collect snow for water can turn into a struggle to survive if a vehicle breaks down and a sudden wind-rise compounds the thickness of the drift. A walk to the dog lines to feed the dogs becomes in such conditions a careful enterprise in which the way is felt along a safety rope from marker pole to marker pole. Vehicles travelling between Scott Base and McMurdo Station or taking scientists on the five mile trip to tend the Beacon satellite equipment at Second Crater carry long ropes. If they lose the road a roped man can walk ahead of the vehicles finding the marker flags but able to return if he becomes lost.

Inside the base the warm round of life continues. As in summer, one man is appointed "house mouse" each day. He must collect snow for water, lay the mess-room tables and wash dishes, bring in stores for the cook, pump up the feeder fuel tanks for the heating plants. During the night watch he takes the meteorological readings at midnight, scrubs out the mess, galley and ablutions hut and at seven o'clock wakes everybody for breakfast. He can also have a bath and wash his clothes, a restricted pleasure in a land in which water is hard won. On Sundays, the cook has a day off and the men take his task in turns.

Amusements are confined to film evenings, party nights and visits, frequently returned, to McMurdo Station. Films are loaned by the United States Navy and shown several times a week. When they have been seen too often it is diverting to show them backwards. Winter is also a time for learning. The men give lectures in their various subjects—astronomy from the surveyor, engines from the engineer—and attend seminar courses run by American staff. Some study for exams to be sat on return to New Zealand. All look forward to their radio telephone talks with home and Radio New Zealand's Sunday night broadcast to the Antarctic which includes messages from families and friends.

On Mid-winter's Day occurs the great Antarctic Festival. A public holiday is declared and climaxed by a banquet demanding weeks of preparation from the cook. The sunset ceremony at which the New Zealand flag was lowered from the flag pole in front of the base seems remote and the first excitement of the day the Sun will return and the flag be raised again is felt.

Antarctica is a do-it-yourself land without corner shops to protect its inhabitants from lack of foresight. Men who winter over must be prepared to turn their hands to almost anything from cutting each others' hair (some do not bother with either hair cuts or beard trims) to making new parts for a disabled tractor. Hobbies and handcrafts flourish. An engineer may have brought an engine block to recondition in his leisure hours. Others paint pictures, build radios, carve wood, knit, or manufacture woollen rugs. The hobbies photographic darkroom has constant winter use while amateur photographers experiment with colour film.

Outdoors it is not always tumult and blizzard. In moonlight the paths between the pressure ridges become still, cold palaces of ice. A wanderer in the night might be amazed to see a seal emerge from a tide crack after some apparently incredible submarine journey up the frozen Sound. In the silence of the night you can hear the pressure-ridge ice creak and snap just as in the snow cave. In the sky there is

46

the endless passage of the stars and sometimes the sudden, violent strain of the aurora.

Throughout man's history the aurora has been known as the greatest wonder of the polar skies. Above Scott Base it takes a common form of bands, rays, arcs and draperies of yellow-green ligh which may last a minute or two or for hours. At times some unexplained contortion of the sky produces many coloured forms—pulsating bands of red and deep green, white rays, blood-red draperies, roseate umbrellas, folds of red and violet, green rays bordered with red. The whole sky may take on a pink glow. At times the sky seems in violent motion. Rays shoot and flicker or light flashes on and off. Bands move across the sky from horizon to horizon. Yet the nature of the aurora is not fully known.

It is apparent to auroral scientists that the Aurora Australis of the south and the Aurora Borealis of the north have individual characteristics even though their simultaneous occurrence in latitudes low enough to allow darkness in the north and south at the same time is frequent enough to be attributed to more than chance. It is clear that aurorae are caused by the same sort of electro-magnetic activity in the ionosphere as is observed by the other branches of upper atmospheric science. But the aurora is the only phenomenon of the upper atmosphere immediately visible to man on the ground and its very obviousness gives impetus to the search for its origin.

The aurora generally has not yet been completely defined even though it has been subject to visual inquiry in the Northern Hemisphere for centuries. Now it is known that auroral light is emitted by excited particles in the ionosphere at a height of about fifty miles and higher. Modern optical instruments analysing the frequency of light waves emitted during auroral excitation can tell the nature of the excited material, suggest the cause of a particular excitement, and measure the heat of the radiant matter. The very accuracy of such instruments has, however, so extended the range of known phenomena which might be called auroral that definition has become increasingly difficult. There is the added difficulty that the aurora takes place in areas shaded from the Sun when other ionospheric phenomenon point to the Sun as their cause. Perhaps if it could then be seen, the aurora might be found to occur in daylight at least as distinctly as at night.

A major problem of auroral definition has been the location of zones of maximum occurrence. Earlier observations seemed to indicate that the zones lay like haloes around the Earth's magnetic dip poles but not centred on the poles. Nor did the "halves" appear to lie opposite each other, as they should if they were governed by the magnetic field. The new concept of the "auroral oval" has replaced the "halo" theory. There appear to be two auroral ovals in the polar night sky, a small display visible in that section of the horizon nearest the Sun (the "day side aurora"), and a much larger display in the sector furthest from the Sun (the "night side aurora"). Because the two ovals would appear in different sectors

47

of the horizon depending on the geographical location of each observing station it was thought that the stations must be seeing part of a "halo" rather than particular displays within the zone of occurrence. Since the I.G.Y. southern auroral studies have also shown that there are several types of aurora occurring inside and outside the zones of maximum display.

Auroral science, like other geophysical disciplines, is now beginning to move out of its descriptive phase and turning towards research on specific problems such as causation. New Zealand has taken a leading part in Southern Hemisphere auroral studies with a collection of observing stations ideally situated on either side of the zone of maximum occurrence. Research is carried on by the Physics and Engineering Laboratory, Auroral Research Station at Lauder. This station scans the clear atmosphere over the high southern plateau country of the South Island. The meridional line of observing stations is continued through Wellington, Christchurch, Slope Point (the southernmost point of the South Island), Campbell Island, Scott Base and the American installations at the South Pole. At times it has been extended north to the Pacific Islands particularly when there has been a possibility of artificial aurorae stimulated by a nuclear bomb test.

In the Antarctic, the standard instrument for recording visual occurrence of the aurora is the all-sky camera, a device consisting of a spherical mirror reflecting an image of all the sky at once which can be photographed on cine film. At Scott Base film records are taken in both colour and black and white. Continuous observation of the position, brightness and colour of the aurora can be maintained without exposing men to the elements. New Zealand's impressive all-sky camera technology was recognised when the country was appointed by S.C.A.R. to supply cameras according to standard specifications for installation in all Antarctic auroral observations and so achieve uniformity in synoptic observations.

The spectrograph is an instrument for recording the components of auroral light. It also has a camera which produces a record on scaled film of the occurrence of each colour in the auroral light spectrum. In an excited state the atoms and molecules of different elements emit different coloured light and thus the spectrograph indicates the presence of oxygen, nitrogen and occasionally hydrogen, tells the temperature of the radiating particles and allows the researcher to guess the cause of their excitement. A photometer is a similar instrument but it is adjusted to select and record the occurrence of a particular colour or wave length of light. These instruments have been used to record the aurora at Scott Base.

Recent research by auroral scientists on the emission of green light from ionised oxygen has shown that this can be seen at all latitudes on all clear nights. In fact, the night sky is not black and lit only by stars, but gains some of its light from the deep forces of the Sun which, though not visible, still affect it. It is thought, in this case, that in the very sparse atmosphere of sixty miles and more above the Earth, atoms which become excited do not immediately release their extra electrical energy by collision with other atoms. There is a lapse of about three-quarters of a second between excitement and collision with its consequent transfer of energy. In the denser atmosphere of lower heights, collision would occur immediately and no light would appear. Science says the higher atoms "hang fire" for this period and so produce their eerie glow.

The causes of the aurora are now being guessed at. Suggestions are that the ionospheric particles are excited by electromagnetic waves passing through the ionosphere or that the polar ionosphere is bombarded by atomic particles energised by the solar wind. It is at any rate certain that the aurora moves across the night sky in time with the changing aspect the revolving Earth presents to the Sun. The aurora also moves in accordance with increases and decreases in the Earth's magnetic field strength and perhaps these motions are caused by passing electric fields associated with the Sun's cloud-like particle emissions.

Auroral studies are only part of the wider field of upper atmosphere research and the interplay of Sun and planet. The data assembled so far from worldwide research suggests a variety of explanations of the shape and function of the space surrounding Earth. It is possible that some of the Earth's lines of magnetic force are directed so far out into space that they can never complete the circuit and return to a conjugate point. The auroral zone may mark the dividing line between those which do return and those which do not. A problem of auroral science here directs the searcher's attention to the condition of the outer limits of the atmosphere and typifies the inter-relationship of branches of upper atmosphere study. Alternatively the Earth's magnetic field may surround the planet like a shell. Inside the shell the lower atmosphere turns with the Earth but the upper atmosphere is fed by the solar wind of excited particles which energises an ionospheric current system concentrating excited material in the auroral zone. Another possiblity is that the same solar winds flow over a region controlled by the Earth's magnetism. The winds act as dynamos generating electric fields which in turn cause currents to flow in the ionosphere, magnetic field changes at the Earth's surface and possibly cause ionospheric particles to emit light. And yet again, it has been suggested that winds in the ionosphere disturb a previously established pattern of ionised particles across the Earth's magnetic field, generating an electric current which maintains the ionisation for longer periods than would occur in calm conditions and thus allowing the particles to emit light as the aurora.

In all these theories, the auroral zones in both hemispheres are seen as the ionospheric areas most affected, and the areas over the poles and within the auroral zones' circular confines as switches which close the electrical circuits set up in the doughnut-shaped shell created by the Earth's magnetic field.

In cosmological terms the division between the day and the night is then entirely arbitrary, even in Antarctica where their opposite natures assault consciousness so forcefully. The solar pattern does not recognise oppositeness but only manifests degree; and this on such an infinite scale that compared to it, mere day and night, light and dark, cold and not-so-cold, isolation and human warmth seem meaningless.

In this predicament, the men who live through the Antarctic winter perform a task which demands resources unextended in the comfortable latitudes of home. Winter is a time of test, and awe, an experience which has made its mark by the time the Sun returns to initiate a chill spring.

7. *The Sea*

THE WATERS OF THE SEA MOVE IN STREAMS ALMOST AS DISTINCT as the flow of rivers through dry land. Antarctica, whose influence upon the motion of waters about the Earth is paramount, drives a series of complex circulations and through them stretches a cold hand to the shores of India and Arabia. This is Antarctic bottom water which sinks down with the freezing of the sea, heavy with salt left when the ice forms, and spreads out like a viscous tide down sea valley and over sea mount, covering half the surface of the planet.

In league with the great hemispheric winds of the south and the turning force derived from the spinning Earth itself, Antarctica has power to move 7,000 million cubic feet of water past the tip of South Africa every second. These seas, where fog swirls thickly all year round, where icebergs drift and crumble and the giant birds wheel endlessly on their sickle pinions, are the richest in the world.

They are part of a distinct body of water called the Southern Ocean which surrounds the polar continent joining the Atlantic, Pacific and Indian Oceans, robbing them of their heat and giving its cold bottom water in return. No other ocean completely encircles the Earth. The Southern Ocean flows eastward without restriction, its narrowest pass being through Drake Passage, 600 miles wide, below South America.

The Southern Ocean's northern boundary fluctuates with the seasons but if latitude 40°S is taken as its limit then it covers 29 million square miles of the Earth. Although this amounts to twenty-two per cent of all the world's oceans it contains only ten per cent of the total heat trapped in the oceanic waters. Because it releases to the Earth's atmosphere three times as much heat as it gains from the sun the Southern Ocean influences the air circulation and thus the weather of the whole planet. Just as the continental land creates the atmosphere's greatest heat sink, the continental waters balance their heat budget by robbing from the waters which surround them.

In the past thirty years research ships of several nations crossing and re-crossing the Southern Ocean have been able to give an approximate account of how it exchanges heat and mass with the atmosphere and other oceans, how it flows across its contorted bed and how it sustains its rich cycle of plant and animal life. Through its research voyages to the Ross Sea and the Southern Ocean south of New Zealand the Oceanographic Institute of D.S.I.R. has made its contribution to this account, building on the foundations laid by the Trans-Atlantic Expedition. Its work is continuing as part of man's search for greater resources to support his rising birth-rate.

Oceanography is another one of the sciences which seek to explain the nature and behaviour of the solid Earth, the atmosphere and the space environment. It is related to the other sciences, needing their assistance to explain the particular phenomena with which it deals and contributing to an explanation of their problems. Tidal movements are a function of gravity, for instance, rainfall a result of air movements over water. The discovery of a glacier-carved valley on the bottom of the Ross Sea is important both to geology and geophysics. The extreme simplicity of the life cycles of animals in the Southern Ocean allows biologists a clearer view of the basic processes of life.

Most of New Zealand's southern oceanographic work is done from the Antarctic supply ship, H.M.N.Z.S. *Endeavour*, the Royal New Zealand Navy's fleet replenishment ship and oil tanker. New Zealand oceanographers sometimes join United States icebreakers for cruises; others work from the land, lowering their instruments through the sea ice of McMurdo Sound.

Their detailed work around Ross Island has included taking photographs of the sea bottom with an underwater camera made by the Institute, collecting bottom samples with a grab, dredge or net and catching fish in traps. They have taken depth soundings, current and tidal measurements and measurements of temperature and salinity of water at various depths. Some of this work has been done from huts built over holes in the sea ice, from ships or simply from the sea ice edge.

Similar measurements have been taken on ocean stations kept on a network of points all over the Ross Sea and meridional courses between New Zealand and Antarctica. Research is now concentrated on the Southern Ocean between the two countries because the biology and hydrology or physical characteristics of the Ross Sea region have been well delineated by New Zealand and previous expeditions.

The area now being studied lies about the Macquarie Rise, a great uplifted area of sea bed south of New Zealand which is a major influence on Southern

Ocean current movements as well as being an important structural feature of the Earth's crust. It demands exploration and understanding for it affects the seismicity of the South Pacific area as well as harbouring information about the tectonic history of the Pacific Basin.

The chief instruments for examining the behaviour of the water mass of the Antarctic or any other ocean are those which measure temperature and salinity, the reversing bottle and the bathythermograph. Reversing bottles are lowered into the sea from the oceanographic ship on a long wire. When they reach a desired depth they automatically turn upside down trapping a sample of water from that depth which can later be analysed for chemical content. An attached thermometer registers the water temperature at the same point. The bathythermograph is a bombshaped device which takes measurements of temperature at a desired depth.

When the depth of water at a station has been discovered by echo sounder the data can be used to draw up a temperature and salinity profile from bottom to surface. Such a profile reveals layers of water of differing temperature and salinity which mix slightly where they adjoin each other but, because of their differing densities, remain distinct. Just as oil floats on water because it is light, so less salty water floats on more salty water.

Measurements of temperature and salinity have gradually delineated the boundaries of the water masses which make up the complex and three-dimensional system of circulation between the Southern and other Oceans.

The masses circulate because they differ in temperature and density and therefore pressure. When temperatures and densities of masses are known the rate at which they must move in their continual effort to equalise pressures can be calculated. Such calculations reveal three major movements in the Southern Ocean.

On the surface of the ocean is the Antarctic surface water, ice-cooled, moving north from the continental shore and, at the same time, moving east in the West Wind Drift caused by the circum-polar winds of the "Roaring Forties", "Furious Fifties" and "Screaming Sixties", as mariners call them. Between latitude 50° and 60° South lies the great oceanographic, meteorological and biological boundary between Antarctica and the rest of the Earth, the Antarctic Convergence.

The Convergence moves north and south seasonally and marks the point where the Antarctic Surface Water, at a temperature just above freezing, meets a Sub-Antarctic Surface Water mass two or three degrees warmer. The Antarctic Surface Water sinks under the warmer mass taking the name of Antarctic Intermediate Water, and moving north. The Sub-Antarctic water mass ends at the next temperature boundary, the Sub-Tropical Convergence at about 40°S. This convergence is the southern limit of Sub-Tropical Surface Water.

Since the northward movement of the surface water must be compensated by a southward flow, a stream of warm Antarctic Deep Water flows south from tropical latitudes and wells up close to the coast to cool and flow north again as the surface

53

water. Beneath these currents the intensely cold Antarctic Bottom Water flows northward along the sea bottom far beyond the Equator. The bottom water sinks because it is cooled to as low as 28.8°F by the freezing of the sea around the coast and because it is heavy with salt precipitated when the ice forms.

Under this circular ocean the Earth's crust is shaped in plateaux, basins, mountains and valleys just as the rest of its surface. The shape of the Antarctic sea bottom has, in fact, provided much information which assists investigation of the nature of the Antarctic continent under its ice sheet. The circularity of the ocean surface is reflected in the nature of the bottom, for apart from the submerged mountain range known as the Scotia Arc, which links the South American Andes with the Grahamland Peninsula, the areas of the Southern Ocean south of the Atlantic, Indian and Pacific Oceans are similar in their topography. They are characterised by plateau-like rises which lie roughly latitudinally across ocean basins.

Scattered round the Southern Ocean are the Antarctic and Sub-Antarctic Islands of volcanic origin, some still active, others almost worn away by wave and wind action. The depth of water over the rises is between 6,000 and 12,000 feet, while the basins lie beneath more than 13,000 feet with occasional ocean deeps of up to 30,000 feet. The volcanic island peaks emerge from shelves about 600 feet deep deepening to ridge tops up to about 6,000 feet deep.

This account of the circulation and bottom topography of the Southern Ocean has only been achieved after many years of careful measurement and calculation not only in Antarctica but on a world-wide scale. If 7,000 million cubic feet a second flow through the channel between Antarctica and Africa then the flow must be fed. The movement is part of the endless cycle of the world's water budget which takes into account the melting of ice, the outflow of rain-fed rivers, evaporation and freezing. On a world basis the budget is far from fully described and studies in the Southern Ocean are crucial to the description.

Among the problems to which oceanographers seek answers are the nature of forces which cause the Antarctic Convergence to take up its seasonal positions; the effect of bottom topography on the flow of currents; the annual and seasonal variations in the quantity of water exchanged between the Northern and Southern Hemisphere; and the theoretical study of the mechanism of current flow around a circular channel such as that occupied by the Southern Ocean. Knowledge of this simplified circular flow will provide criteria for measuring the more complex circular movements of water masses around the world's great ocean basins.

If long-range weather forecasting is to become possible for inhabited lands, such annual and seasonal ocean movements will have to be known. If ocean movements become predictable then their effects on weather will be predictable.

The circulation of the Southern Ocean also has a decisive influence on the quantity of living organisms which it supports. The Antarctic surface water provides ideal conditions for the growth of a vast quantity of phyto-plankton or microscopic

54

vegetable life. With the aid of sunlight these organisms convert nutrient salts such as nitrates and phosphates directly into living cells by the same photo-synthetic process used by land plants. In the long hours of summer sunlight great quantities of phyto-plankton of a variety called diatoms grow or "bloom," sometimes so thickly that the sea in the Southern Ocean is stained brown. The supply of nutrient salts in the surface water is constantly replenished by the upwelling close to the continent of the Antarctic Deep Current which constantly bears southwards salts released in tropical latitudes by the bacterial decay of vast quantities of dead organisms.

While the number of species of diatoms and of the zoo- or animal-plankton which feed on them is limited their life spans are extended by the cold water so that many more generations co-exist. It is this combination of constant replenishment of nutrients, long periods of sunlight and longevity which makes the Southern Ocean richer in life than any other in the world. Just as temperate latitude land animals are ultimately dependent on green plants for their food-chain cycles, so the life cycles of Antarctic animals without exception depend on the profuse production of the Southern Ocean's diatom "pastures".

In summer, the plankton bloom continues all the way north to the Antarctic Convergence where it diminishes abruptly with the change in water temperature and along with it the high density of animal and bird life. Some zoo-plankton take advantage of the Antarctic Deep Current, passing their developmental stages drifting south with it to emerge as adolescents near the coast in spring and spend their adult summer travelling north to the Convergence there to reproduce and continue the cycle. The easterly drift of the Surface Water also insures that both zoo- and phyto-plankton species are evenly spread around the whole continent. The distribution of the higher animals is inclined to reflect this condition.

Because the Southern Oceanic waters contain little calcium, the diatoms use silica to construct their tiny skeletal forms. Bottom samples taken by oceanographers from the Southern Ocean bed show that its top layer is composed of diatomaceous ooze which lies in a belt right round the continent and is built up by a constant rain of dead diatoms sinking to the bottom. North of the Antarctic Convergence, the ooze is not found on the surface, but sea bed cores have shown that under sediments typical of the Sub-Antarctic seas lies a layer of diatomaceous sediments. The sediments indicate a period when the Earth was colder, the Antarctic ice cap bigger and the Convergence lay further north.

The physical characteristics of the Southern Ocean have, then, great importance for biological life generally and in particular for man, who is now considering ways in which the richness of the sea can be used to feed an ever-growing world population. Any attempt at large scale utilisation of the southern food resources will demand a complete and detailed knowledge both of the physical behaviour of the sea and the animal life cycles which take place in it. Antarctic oceanography is the basic study behind any such practical developments.

SOUTH

To cross the Antarctic Convergence travelling south by ship is often to make a dramatic entrance to the polar world. The change in water temperature of only two degrees is sufficient to have a considerable cooling influence on the atmosphere. North of the line at which the Antarctic and Sub-Antarctic waters meet in summer the air will still be fresh and spring-like, but to cross the line is to feel the chill of ice. Suddenly there are sea birds wheeling and turning in the almost constant mist caused by condensation of moisture in the warmer northern air as it moves over frigid water. Icebergs are more frequent and a few hundred miles south the edge of the pack ice straggles across the ship's course. Antarctica presents another paradox, that of profuse life in apparently sterile cold.

In its other aspect, the Southern Ocean is the most turbulent in the world. Without land to halt them, the great cyclonic southern storms move ceaselessly around the world. The ocean surface is never still from New Zealand south to the edge of the pack ice. In the "Furious Fifties" and "Screaming Sixties" sixty-knot winds are common, and with them seas which tower up to sixty feet high above the ships which roll their way violently to and from the continent in summer.

In the Ross Sea itself, the land gives some protection from the giant Southern Ocean swell but winds are seldom still. Ships have the added hazard of ice and sometimes sail surrounded by hundreds of icebergs seen only by the electronic eyes of their radar equipment. In heavy weather spray freezes on their superstructure sometimes covering it with hundreds of tons of ice.

In such conditions, Antarctica's oceanographers must carry out their shipboard work. Days of complete calm are rare even far south on the Ross Sea. At each ocean station the research ship must stop and lie drifting broadside to the swell with several thousand feet of wire trailing instruments into the sea depths. Since the ship drifts down-wind the work must be done on the windward side exposing men to wind and chilly spray. With cold fingers they must sift the mud, weeds and animals hauled from the bottom by dredge, grab, and coring drill, carefully searching for life and the remains of life which tell the nature of bottom conditions. Under way they tow fine nets through the surface water collecting delicate planktonic life which must be carefully separated and preserved by fingers clumsy with cold.

On the bridge the echo sounder endlessly charts the sea depth or reveals a bottom shape of peaks and valleys. The bouncing radio wave sometimes receives a shadowy echo from rich layers of living plankton, for under its moving and broken load of ice the Southern Ocean heaves with life.

8. *The Animals*

No OTHER SOUND OF THE ANTARCTIC SUMMER IS AWAITED with such longing as the sound of waves breaking on a shore. It is the sound of freedom to feed, to breathe, to be born, to grow. It is the sound of the sea's release from ice with all its consequence of rich and varied life. Penguins need no longer trudge a dozen miles or more across the sea ice from their rookeries to fetch food. Seals need no longer grind their teeth on the ice to carve their breathing holes. Flying birds can fish the sea close to the restful land.

In the sea itself the thick bloom of plankton can begin again renewed, perhaps, by the release of organisms which have lain frozen and dormant all winter in the ice. In Antarctica all living things owe their vitality to the sea, from simple bacteria to the supremely developed Emperor penguin. Lichens hundreds of miles from the sea and thousands of feet high among mountains find sustenance in the wind-blown manure of coastal animals.

Minute insects browsing among the sparse lichens suspend their life processes when conditions are unfavourable and exemplify the unparalleled simplicity of the Antarctic biological world. They mark the end of the longest Antarctic food chain based on the summer bloom of diatoms in the sea. On diatoms the shrimp-like zoo-plankton, krill, feed. Fish and squid eat krill. Whales, seals, penguins and flying birds eat squid and krill. Bird manure feeds land and freshwater plants and the insects which eat the plants feed nothing. Thus Antarctica presents biology with one of its most clearly defined examples of the inter-dependence of living forms.

Stark and simple in its superficial appearance, Antarctica was long thought of as a sterile land. Scientific expeditions of the first decade of this century carried out a quantity of research in the life sciences but until recent years the physical aspects of the land, sea and atmosphere have commanded a far greater scientific effort. Life scientists now realise the outstanding importance of research in an environment uncontaminated by man and are turning attention to the urgent need to pursue descriptive biological studies in Antarctica and attempt solutions to the problems which are revealed.

One result of this new recognition has been that scientists of the Antarctic nations working through S.C.A.R. have laid down the general principles for an international agreement on the conservation of nature in Antarctica.

They state that Antarctic plants and animals are of outstanding scientific importance and that man's influence on them should be minimal. Many Antarctic life forms are unparalleled in the world, display perfect adaptation to an extreme environment and can be successfully studied only in their natural state.

Alone among the world's great land masses Antarctica has been little affected by man and should be preserved intact as a world heritage. Antarctic plants and animals are particularly vulnerable to interference because they are so highly specialised that they cannot readily re-adapt to an altered environment. Marine animals which breed on land, such as seals and penguins are defenceless for instance, against unaccustomed land predators. Because the animals and plants depend directly on the sea and its store of planktonic food any interference with the biological system can have immediate and far-reaching repercussions.

The world's scientific and economic needs make justifiable demands on Antarctic plants and animals. Because these demands will increase, biological research must take place before Antarctic life is utilised in any new way. In fact, man must learn how to manage the wildlife without doing it unnecessary harm, and must prevent pollution and contamination of Antarctic air, land, water and ice. Some Antarctic animals are restricted in their range, others have a circumpolar range and many migrate seasonally to other parts of the world. Responsibility to conserve Antarctic life thus extends far beyond Antarctic and national boundaries and demands international co-operation on a large scale.

Conservation such as the scientists propose and even eventual management of the Antarctic life complex to help supply world needs for food, drugs and raw materials will be one of the most challenging endeavours that man has undertaken. Such goals could not be reached without a complete knowledge of food chains, behaviour of animals during various stages of their life cycles, and the physical means which they have developed in order to survive in a harsh world. A deeper search into the relationships between the animals and the thriving microscopic undercurrent of their environment would also be essential.

The fact, for instance, that microscopic diatoms produce antibiotic substances which inhibit the growth of parasitic bacteria has wide repercussions. Antarctic birds and seals which feed on diatom-fed krill are in turn supplied with antibiotic substances which in some cases make their intestines bacteriologically sterile. Research is necessary to discover if the diatoms' facility to manufacture antibiotics

58

can be used by man as a supply source of drugs which fight his own diseases. On the other hand it is not known if diatoms and other phyto-plankton produce poisonous substances which might be harmful to man should he utilise other life forms in the planktonic food chains. It is not known if any such substances are poisonous to fish. Management of a food source implies providing optimum conditions for production and such facts would have to be understood if the sea were to be "farmed".

Similarly little is known about the microbial relationships between Antarctic animals and man—whether or not organisms harmful to either might be exchanged through contact at various levels. Whales are thought to be susceptible to tuberculosis, and penguins to carry viruses which cause respiratory infections and other more serious though rare diseases in man. Increased human contact with Antarctica and its animals might bring public health problems for both worlds. Skua gulls show some tendency to forsake their natural foods—fish and krill, carrion and penguin eggs and chicks—and live on the refuse of Antarctic stations. It is possible that they are being exposed to pathogens or disease-causing organisms which would be redistributed around the world through transmission to other animals. Here is another reason why man's Antarctic endeavours so far commit him to growing involvement and research.

Viruses which cause dysentery in man are known to exist in penguin colonies. Tetanus spores have been found still viable in pony droppings from Scott's last expedition. Antarctica in the summer is not so cold that unpreserved organic foods fail to decompose and it is on the slow processes of bacterial decomposition and return of nutrient chemicals to continuing life cycles that the Antarctic life depends just as life depends in other lands.

The environmental processes are so simplified however, that outstanding opportunities to investigate the sources of life are available. It is possible that microbes might travel through space on meteorites and if any life was found on meteorites which fell on the uncontaminated Antarctic ice sheet it would be very easy to determine whether or not it was of extra-terrestrial origin. Deep ice cores will tell if viable microbes can exist in the depths of the ice sheet itself. Little is known about the process which allows plants called algae to exist in glacier ice or to bloom as a red stain in the rotting summer snow of coastal fields.

The unknowns of Antarctic biology are just as manifold when it comes to research in all aspects of higher life, from the sponges and starfish which swarm on the bottom of McMurdo Sound to the heat-conserving mechanisms of seals and penguins. What specialisations in breeding behaviour allow Antarctic petrels to sustain bird populations among the world's largest in the world's most rigorous environment? Here are mysteries which will take years of patient field observation and laboratory examination to elucidate.

While the abrasive action of recurrent ice prevents the growth of any inter-tidal and shore zone life in the Antarctic sea, coastal sea bottoms are rich in animal life. Sponges are in such abundance that biologists have been led to conclude that they have reached their evolutionary apogee in the Antarctic environment.

Most bottom animals are, like sponges, of the filter-feeding type; that is, they collect their food by sifting organisms from the plankton-rich water rather than feeding on organic detritus washed down from the inter-tidal zone. In some areas, however, the bottom animals are characterised by shellfish and the starfish which feed on them. Fishing is not a profitable pastime in McMurdo Sound unless traps are used. Several species of fish, mainly nototheniiforms, are found however, some of small size and some giants between 40 and 150 pounds. These largely support the coastal Weddell seal populations.

Just as the small number of Southern Ocean plankton species is compensated for by the fact that individual lives are longer and more generations exist together at once, the higher animals too are characterised by great numbers and a lack of diversity of species. The largest animals, the whales, have so far provided the only means whereby man can tap the food resources of the Southern Ocean. The effects on whale populations of the whaling industry have been dramatic evidence of the need for conservation, and some progress such as the international agreement on the total protection of the right whale has been made. The whaling nations also agree to a quota system; but the conservation of commercial whale stocks has remained a recurrent problem in the face of serious decline of such varieties as the blue whale, the largest whale of all.

Men on the ice have little contact with whales, for the great commercial grounds lie far out at sea. When the sea ice breaks up close to the land killer whales often appear on their hunt for seals. They have not been a subject for biological research and yield insufficient oil to be of commercial value but remain the most feared creatures of the pack ice. Explorers have testified to apparent attempts by killer whales to tip them off ice floes and have even maintained that they display intelligence in their efforts to dislodge seals from floes, bumping the ice with their heads from underneath in an apparent attempt to split it.

Three of the four types of Antarctic seal are creatures of the pack ice ranging to the pack limits far from land. They are the crabeater seal, the rare Ross seal and the sea leopard or leopard seal. The Weddell seal is a true creature of the Antarctic shoreline and lives close to the land throughout the year, keeping its breathing holes in the sea ice open by gnawing with its strong teeth. Crabeater seals feed on krill, Ross seals, it is believed, on squid, leopard seals on penguins and Weddell seals on fish. Other seals which visit Antarctica but are normally found in the Sub-Antarctic are the elephant seal and fur seal which became almost extinct because of commercial killing by the nineteenth century sealers.

Weddell seals breed round the continental shores, dropping their fast-growing pups directly on the ice beside the melt holes or tide cracks through which they can enter the sea. McMurdo Sound has a Weddell seal population of several thousands, a few hundred of which spend the summer on the ice about the pressure ridges at Scott Base. Their pups arrive in November and "baa" like lambs. Crabeater and leopard seals are seldom seen in McMurdo Sound and Ross seals appear never to leave the pack ice.

Of the thirty Antarctic region birds, only fifteen species including penguins nest on the continent itself. While there is a large food supply available suitable nesting sites are few and therefore, in many cases, shared. Skua rookeries adjoin penguin rookeries, some of the skuas benefiting by preying on the penguins. Small species of petrel share the rocky cliff faces where they nest in burrows and crannies. Some petrels nest hundreds of miles inland because of the shortage of coastal sites.

Birds or the petrel and tern families feed on krill, but some like the tiny snow petoel or "elegant white petrel" as Sir James Clark Ross called it in his logbook, eat surface fish. In the extreme mainland environment, there is a high mortality of both eggs and chicks, but large populations are sustained because mortality from other than natural causes is low among the ocean-ranging adults. A single egg produced once a year by monogamous parents who use the same nest site each year is adequate to maintain populations among the highest in the bird world.

Of the four species of penguin which breed below the Antarctic Circle only two can be considered true Antarctic penguins of circum-polar distribution. These are the Emperor penguin, largest of the family and the Adelie, much smaller, the most beloved and best known Antarctic animal. The other two are the gentoo, which breeds in some Sub-Antarctic islands and on the Antarctic Peninsula; and the chinstrap, long thought to be restricted to the South American sector of Antarctica but recently found nesting with Adelie penguins in north Victoria Land rookeries.

The circum-polar spread of the chinstrap penguin is of great interest to biologists and some have suggested that it may be connected with the decline in whale populations and the consequent increase in available quantities of the zoo-plankton, krill. Except for the Emperor which feeds on squid and fish, Antarctic penguins are chiefly krill feeders. Adelies have been known to eat fish at times when pack ice about their rookeries was too heavy to allow a plentiful growth of plankton. They are all marine animals, well adapted to live at sea for most of their adult lives, coming ashore only to breed and moult.

Penguins are also limited in their choice of breeding sites for they must be able either to scramble ashore to the rookery through surf or to jump from the water onto the ice edge. Consequently populations at a few breeding sites are large. At Cape Adare half a million Adelies congregate each summer with far smaller populations at less than a dozen other breeding places on the Victoria Land coast, Ross and smaller islands.

Emperor penguins, which represent the highest evolutionary development of the penguin form, are even harder pressed for breeding space. Because they are big birds (up to eighty or ninety pounds) their chicks must be born early enough in the year to ensure they grow sufficiently mature to survive the following winter. Consequently the Emperors breed in early winter and incubate their eggs through the most intense cold of the year on sea ice rookeries which after float away with the summer breakout. Male birds incubate the eggs, holding them on the top of their feet tucked under a downy skin flap, and fasting for up to three months during the process. This allows the female to be ready with food brought from the open sea which is often many miles away when the eggs are hatched. Mortality of eggs and chicks in this most difficult breeding cycle is high. Little more than a handful of Emperor penguin rookeries are known on the whole Antarctic coastline yet the species manages to maintain itself with a success that fits the dignity of a kingly bird.

Penguin species exhibit similar types of breeding behaviour, usually congregating in rookeries which allow protection from the elements and predators, and a satisfactory system of mate choosing. None live their rookery lives with as much delightful domesticity as the Adelies and it is no wonder that their behaviour has been studied with such great interest from the time of earliest exploration.

The first Adelies return to their rookeries on about 16 October each year. They toddle along the sea ice singly, in groups of three or four and sometimes in long straggling lines consisting of hundreds of birds. At the rookery mates recognise each other by courting displays—curious dances accompanied by snake-like neck waving and raucous cries. Male birds build a nest of a ring of stones on top of a nest mound built up by years of guano deposition. Female birds accept the mate and the nest, assist in the completion of building and when the two eggs are laid leave the male to incubate them while they return to sea to catch food for the chicks' first meal. The first chicks are hatched in mid-December and at the age of about four weeks gather in groups called crèches, seeking mutual protection from the weather and the preying skuas. At the end of January chicks have lost their down and begin to leave the rookery. They swim north to the pack ice where they spend the winter.

Because of their tameness and gregariousness in the rookery Adelies are particularly valuable research subjects for the biologist, and the growing knowledge of their relationships with their environment, behaviour and physiology will contribute to the solution of wider zoological problems. Their behaviour continually fascinates the observer, who can easily be led into errors of understanding through attributing to the birds "human" qualities because of their quaint and characteristic movements and expressions. In fact, the Adelie is a markedly unintelligent bird although he displays considerable cunning in the theft of nest stones from his neighbours and fighting tactics during the ensuing battles.

On land skua gulls are the only predator animals affecting the penguins. Skua

THE ANIMALS

rookeries are normally associated with penguin rookeries but the skuas do not feed indiscriminately on penguin eggs and chicks. They divide up the penguin rookery into hunting territories which are jealously defended by a few pairs while the remaining skuas are forced to feed at sea. Because of skua predation, human disturbance of penguin rookeries by careless movement through the sitting birds, or through disturbance by vehicles and aircraft, can have drastic results. Adelies desert their nests when they are frightened leaving eggs and chicks to the skuas which, though they display a certain cowardice in the face of determined penguin defence, ruthlessly attack unguarded nests.

Human influence on the rookery environment can disturb the penguins in other ways. Cape Hallett Station shares Seabee Hook in summer with a rookery of about 160,000 Adelies. Buildings and stores left lying on the rookery area cause deep snow drifts where the surface was previously swept bare of snow. In the thaw the snow melts to form pools which inundate penguin nests. Individual penguins tend to use the same nesting sites year after year without deviation, and so lose their eggs in the floods.

New Zealand's Antarctic biological research has been carried on by the Trans-Antarctic Expedition, Antarctic Dvision; the New Zealand Oceanographic Institute (in marine biology); the Dominion Museum; the Antarctic Biological Unit of the University of Canterbury; the Victoria University of Wellington, and Otago University. Their studies in Antarctica have made an important contribution to the description of the relation between land, sea, animals, and plants in the Ross Sea region. Some New Zealanders have contributed to the United States Antarctic Research Programme in summer field work, at times travelling with New Zealand parties but contributing their research to American scientific organizations.

A biologist is sometimes included in the wintering-over staff at Scott Base. A small biological laboratory is used for the study of material collected during the previous summer, and winter studies such as on the general physiology of Husky dogs are carried out. The chief study conducted from base has been on the McMurdo Sound Weddell seal population and breeding cycle of which little is known. Aerial photography of seal rookery areas through the summer is done in conjunction with the ice breakout photography for the glaciological programme, and reveals fluctuations in population. The resident biologist also supervises the taking of seals for dog food, a necessary practice carried out according to conservation principles.

Penguin research has been concentrated at Cape Hallett, Cape Bird, and Cape Royds on Ross Island which, with about 1,000 pairs, is the smallest and the farthest south Adelie penguin colony. Population surveys have been carried out from time to time on the Cape Crozier Emperor penguin colony and the Adelie colonies at Cape Crozier, Beaufort Island, and Cape Adare. Research interests include periodic egg, chick and total population counts during the breeding season, banding of

63

adult birds to discover their range, parasitology, weighing and measuring of eggs and chicks, measurement of body temperatures, general observations of behaviour of both penguins and predating skuas, and experiments in mate, chick, and egg recognition by individual birds.

The Cape Royds colony where the University of Canterbury work was first concentrated is particularly interesting for here, at the southernmost point of their range, the Adelies live in extreme conditions. That the Cape Royds rookery was once much larger is suggested by soil analyses by staff of the D.S.I.R. Soil Bureau who have found guano remains over an area considerably large than the present site. A particular aspect of research at Cape Royds concerns the effect on the small rookery of frequent summer visits by men and machines. During the first few research seasons at Cape Royds, biologists lived in the hut built by Shackleton's British Antarctic Expedition of 1907–09. Now the university biologists concentrate their research at Cape Bird where the summer huts of Harrison Station have been built and a larger rookery affords greater opportunities for research.

Now all Antarctic biologists are turning from general description to attempts to solve particular problems. For instance, cold-adapted fishes may convert food energy into growth more efficiently than those in warm waters. When such a mechanism becomes understood through a study of Antarctic fish a way might be found to apply its principles to commercial fisheries.

Many Antarctic birds are migratory and thus play their part in the dispersion of animal life around the world by carrying invertebrate animals and micro-organisms from place to place. Their ability to spread life in this way demands documentation. Little is known about how long Antarctic seals live and the effect on their adaptation to cold and deep water; or about their orientation and navigation senses under ice and water. Such problems are now gaining increasing attention from biology.

Nobody knows why Antarctic animals should sometimes wander far from the sea. Skua gulls have been seen on the Polar Plateau south of the Beardmore Glacier and quite close to the Pole. Penguin tracks have been seen in melt water mud beside the Koettlitz Glacier moraines twenty miles from McMurdo Sound. The carcass of a seal, mummified by the cold, has been found 6,000 feet up on the slopes of Mt Discovery and many other mummified seals and penguins in the dry valleys of the McMurdo Oasis. Perhaps they lost the sea and wandered inland until they died of starvation. Perhaps they were driven by some instinct to die on the land. Because they were alive they must have had some purpose which nobody now knows.

9. *The Journeys*

SLEDGING THERE IS NO SOUND BUT THE UNEVEN DRUMMING OF the sledge runners, the dog's crunching footsteps and the long sheer of skis on dry and sandy snow. The Polar Plateau curves down a far horizon like the ocean. The nine dogs scoop mouthfuls of snow as they run, leaving bright blood stains from their cut lips. The runners drum while the two men grip the sledge handles and tow ropes for balance, and ski with a long, gliding walk. They shout encouragement to the dogs ("*Huit* Apolotok! *Huit* Peabrain! *Huit* Suzie!"), their voices thin in the high, thin Plateau air.

On the Plateau there is no gauge for time but hunger, and cold in the bones. Distance seems as meaningless as a dream of home and the measurement of miles by revolutions of a bicycle wheel behind the sledge pointlessly finite. But a day's progress continues—two, three, even five miles an hour on good, fast snow. When it is time to rest, one skier kicks his boot free from the loose ski binding and brakes the sledge as he calls "Aaaaah boys, aaaah boys," in a soft and lowered voice. The dogs halt, flop on the snow and scoop up great mouthfuls, for it is their only water.

Sledging has its own lore and philosophy. It is an art handed on by successive field party men as the journeys of Antarctic exploration continue. The dogs provide it with a living continuity, learning to pull a sledge in their adolescent summer, hauling thousands of miles through their seasons of prime to end their lives helping to train their pups in the traces. Likewise, the sledging man must learn his craft from the intricate lashing of a hickory sledge to the stitching of wounds after a dogfight. He must learn endurance to work out the bitter hours of an alpine survey station and patience to bear incarceration in his tent throughout a week-long blizzard.

Little of Antarctica now remains unseen for most of the continent has been at

65

least flown over. Indeed all the 90,000 square miles which New Zealanders have mapped have been previously photographed from the air. Photography, however, does not reveal the real nature of the land but only the relative positions of its most prominent points. The men who sledge or walk over the snow are still explorers in a real sense, for all the mysteries of the landscape are revealed as they record their experience of it in surveyors' sight-books, sketches and geological notes.

Their maps supply the framework on which the data from all other scientific enterprises in Antarctica can be plotted. Geology, seismicity, gravity and magnetic studies, the incidence of animal populations and particular species, meteorology, the positioning of the aurora in the night sky, the siting of tracking stations for Earth satellites, the charting of safe routes for land, sea and air travel—all depend on accurate maps and the correct positioning on the Earth's surface of stations where observations have been made.

Antarctic cartography is just as much an international activity under the surveillance of S.C.A.R. as any other Antarctic endeavour, and New Zealand's contribution is the production of maps of the Ross Dependency and the Polar Plateau areas bordering the Trans-Antarctic Range. Production standards for finished maps conform to a S.C.A.R. agreement on accuracy, scale and colouring. Such maps can be used as a spatial reference for any scientific observations as well as a starting point for geodetic measurements or those concerned with the shape of the Earth as a whole.

Topographical and geological maps are an essential part of the whole complex of Earth sciences and symbolise the work of describing the planetary environment in an immediate way. They tell the present state of the surface of the Earth, the superficial cover which hides a continuing process of movement and change in the Earth's crust and beneath it, the causes of which can only at present be surmised. Because of Antarctica's key role in explaining such problems as continental drift, the origin of earthquakes, and the effects of glaciation, mapping has been one of New Zealand's biggest Antarctic commitments. Political considerations have also made it expedient for New Zealand to have a complete knowledge of the territory claimed in Antarctica even though such claims are now suspended under the Antarctic Treaty. The act of exploration has traditionally been related to the act of claiming territories in a new land.

In general the cartographic coverage of Antarctica achieved by all nations is only just growing to reasonable proportions. Portions of the continental coastline are still unexplored however, and may be inaccurately positioned on the sketchy maps available. In most cases, mapping which has been completed has been restricted to coastal areas which allowed easy access.

Like most Antarctic cartography, New Zealand's maps could not be produced without the aid of high altitude oblique aerial photographs. United States Navy

66

reconnaissance aircraft supply these for New Zealand. Such photography is not sufficient in itself for the production of maps because an accurate ground control network of triangulation is necessary to give the correct latitude and longitude of features shown in the photograph.

For New Zealanders, the dog team and sledge was the favoured method of travel through survey areas while ground control data was obtained and local geology surveyed. Most of the Ross Dependency Survey has been completed by field parties organised by Antarctic Division, D.S.I.R. The standard pattern for field operations was to send two parties, each with four men and two dog teams, into the field for about four months in the summer. At least one surveyor and one geologist accompanied each party. At the beginning of New Zealand's Antarctic field work during the summer before the I.G.Y. began, the first reconnaissance journeys were made by a party with man-hauling sledges similar to those used by the early explorers. Man-hauling was used for several journeys such as the Tucker Glacier survey expedition to North Victoria Land by a New Zealand Geological Survey party, various Victoria University of Wellington expeditions and those by the New Zealand Alpine Club and the Federated Mountain Clubs of New Zealand, privately organised ventures contributing to the Ross Dependency Survey and operating with the assistance and surveillance of Antarctic Division. But now motor toboggans, small vehicles of higher haulage ability than a good dog team are used for small-scale expeditions of a more specific nature.

Dog sledging remains the most adventurous, however. New Zealanders have driven dogs in the Trans-Antarctic Range to higher altitudes than ever before recorded and with their help have climbed the highest peak yet scaled on the continent, Mt Fridtjof Nansen, which towers 13,700 feet over Scott's route to the Pole, the Beardmore Glacier. In journeys covering up to 800 miles in a summer they have traversed the mountains and glaciers from the Admiralty Range in the north, Ross's first sight of the continent, to the Queen Maud Mountains only 300 miles from the Pole. One party emulated the feat of the discoverer of the Pole, Roald Amundsen, by sledging down the gigantic ice falls of the Axel Heiberg Glacier south of the Beardmore on the fiftieth anniversary of the Pole's discovery.

New Zealand's sledging men have rarely been professional explorers, Antarctic or otherwise. Normally, the surveyors are staff members of the Lands and Survey Department of New Zealand, which has an Antarctic division attached to its cartographic branch and is engaged solely in the production of Antarctic maps. Geologists have been provided by the New Zealand Geological Survey and the universities, and produce their own geological maps based on surveyed topography. Occasionally both surveyors and geologists from other countries successfully apply for positions with New Zealand expeditions. Field assistants come from many occupational groups but are always experienced alpinists, sometimes veterans of Himalayan expeditions as well as of the splendid climbing country of New Zealand's

67

SOUTH

Southern Alps.

This fund of alpine experience is also useful to the United States for in recent summers a group of New Zealand climbers has provided mountain- and snow-craft instruction courses at McMurdo Sound for United States Antarctic Research Programme field party members about to set off on summer expeditions. This is one of the services provided by New Zealand in return for air support for field operations given by the United States Navy's Air Development Squadron Six, the squadron responsible on the continent for all air support for Operation Deep Freeze using C130 Hercules aircraft. Assistance to New Zealand includes re-connaissance flights for field party staff, placement of field parties with their sledges and equipment in survey areas sometimes hundreds of miles north and south of McMurdo Sound, and helicopter transport of foot-travelling geologists to the Dry Valleys or biologists to the penguin rookeries of Cape Bird and Cape Crozier on Ross Island. Ski-equipped Douglas Dakota aircraft historically formed the basic transport fleet for field party support. Helicopters assist smaller groups close to the Sound. Recent operations like the establishment of Vanda Station have been assisted by R.N.Z.A.F. Hercules aircraft.

Field parties in the Trans-Antarctic Range for up to four months at a time must be re-supplied at least once during the season. The Navy pilots have to find camp sites in un-mapped areas, make parachute drops of man rations and dog pemmican, or land on improvised snow airstrips with unpredictable surfaces. They must take off again struggling for lift in the thin air of altitudes up 10,000 feet even with the rocket thrust of "jet assisted take off" bottles. Every New Zealander who has sledged in Antarctica has a warm gratitude and admiration for the Americans who helped his party.

The surveyor's task is probably the coldest that Antarctica can offer. A major survey station on top of a prominent mountain peak or nunatak may have to be occupied for as long as twelve hours before a complete round of observations is complete. The position of the point must be fixed by taking sights of the Sun and even of the stars, which are faintly visible if the surveyor knows where to find them. Then the angle of bearing of every other topographical feature must be read with the theodolite. A photographic panorama of the landscape right round the station is taken on 35 mm film and a pencil sketch panorama is also taken to identify the points photographed.

When the Ross Dependency maps were drawn, back in Wellington, aerial oblique photographs taken by a Neptune reconnaissance aircraft flying at about 25,000 feet were used to provide topographical details such as glacial flow lines, ice patterns and rock outcrops. The same photographs help the geologist when he comes to superimpose his map of the survey area's geological formation on the topographical outline fixed by the surveyor's pattern of triangulation.

The naming of features discovered and surveyed is one of the pastimes of field

68

party living. Most names can be decided in the field according to rules laid down by the New Zealand Board of Geographic Names. The rules ensure a standardised system of reference and description of features, and also that the importance of a feature is matched by the dignity of its name. A mountain range for instance, would not be named after a sledge dog but perhaps after an expedition leader or patron in New Zealand. Sledge dogs, in fact, are seldom commemorated although some have given such service that their names have been, as it were, immortalised in reference to minor features.

In spite of the development of modern motor toboggans and heavier oversnow vehicles, dogs will always have a place in Antarctic travel. They are slow-but-sure transport and they can pull sledges in country a toboggan could not tackle or where heavier machines would not be safe. Because they have hearts, blood and muscles they are noisy and companionable, tireless and unrelenting, and will pull until they fall dead in the trace. If their driver is starving he can eat them.

Huskies in the nine-dog team favoured by New Zealanders pull sledge loads of about 1,100 pounds on field journeys while motor toboggans, with their bigger sledges manage about 1,800 pounds of fuel, food, tentage, survey equipment and geologists' rock specimens. Weight-for-weight the quantities of dog food and fuel consumed daily are in favour of the motor and, since it tows more, its range before re-supply is greater. But a motor toboggan cannot be driven on short rations and dog teams do not have mechanical breakdowns. There is hardly a man who has worked with them who would swap "shaggy dogs" for "tin dogs", as the motor toboggans are called, even though dogs demand more attention and hard work in care and driving. Toboggans have proved more efficient however, for small-scale expeditions and journeys close to Scott Base such as those connected with the McMurdo Ice Shelf glaciological study. They have the further advantage that they need no maintenance during the long winter which follows a short sledging season while dogs must be fed, exercised and kept healthy.

Sledging begins, then, with the breeding, rearing and training of Husky dogs, the famous draught dogs of the polar north, equally efficient in the south. Wonderfully adapted to Arctic living, the Husky has faced the harsher Antarctic with few signs of discomfort. When the temperature rises to about freezing he is too hot and goes miserably to work. When a blizzard blows he curls up in a ball to be buried in snow. Occasionally, in the depth of winter cold, he becomes frostbitten but lives out the night in the open to greet the first training runs of spring with energy and vigour.

Of the sixty or so dogs at Scott Base during the peak of exploration work only about twenty-five remain. They came originally from Greenland, from the Australian Antarctic Base, Mawson, and from the Auckland Zoo. The strain has been improved with fresh Greenland blood and breeding is controlled as much as possible in an effort to produce the best type of dog. The ideal is a thick-set animal, the male weighing about a hundred pounds, with thick, short hair. Long-haired puppies must be destroyed as they could never survive the winter. Snow would mat in their hair so that it froze to the ice.

Pups are raised in a "maternity home" close to the base buildings. They arrive haphazardly at different times of the year but winter mortality can be high as some Huskies fail to make good mothers. Fed on milk, baby food and vitamins, then minced seal meat or steak, they grow astonishingly fast and soon romp through the base much petted by the men. Childhood ends early for Huskies, however. At four to six months old they start to fight, steal food and run away as far as McMurdo Station. They are tethered out with the teams and by the end of the first year are learning to pull a sledge with the base team or, in the case of a bold, intelligent dog, to lead.

Huskies are pack animals. Each team develops its own organisation and social ladder. They fight viciously among themselves for precedence or the favours of a bitch. The team leader may not be the strongest dog or the best fighter, for the desirable qualities in a leader are the ability to run straight and obey commands to start, stop and change direction. Although they fight, Huskies are the friendliest dogs to their drivers, particularly at feeding time. Field rations allow them a pound-and-a-half of pemmican, a vitamin-fortified meat-meal block, daily. When they are spanned out in teams at the base dog lines they are fed six pounds of seal meat or its equivalent food value of pemmican every second day.

The Husky's greatest pleasure is to pull a sledge. When it is time to break camp in the field and harness up for the day's run they howl their excitement. When half the team has been hitched to the picketed sledge they will quite likely start to fight while their fellows are being harnessed. The drivers must kick the snarling jumble of dogs into quietness, punching them with leather-mittened fists, beating them with rope ends before they wound each other, for on the Polar Plateau a badly wounded dog is as good as dead.

A team starts pulling with a wild rush of speed over the first two or three hundred yards. Then they settle to a quick and steady walk which they will continue for hours until the lunchtime halt when the men drink coffee brewed at breakfast time and kept hot in a thermos, eat a bar of chocolate and a slab of fruit cake—rich, sweet food for heat-producing energy. The halt will not be long, for sledging is hot work and a lightly clad body cools quickly. A cigarette tastes harsh and dry in the cold and can only be half-smoked otherwise the butt would burn the clumsy, heavy mittens that are called "nose wipers" because they have a woolly back to brush away the ice of frozen breath.

The two teams in a party keep well away from each other to avoid exciting the dogs. They follow a course with a sun compass, somewhat like a sundial, for the Magnetic Pole is too close for a magnetic compass to be accurately used. Their course may be a slow drag up a steep slope with one man in front hauling on the lead trace while the other tries to stop a thousand pounds of sledge and load from falling back; or down an icy slope when rope brakes must be tied around the sledge to increase the friction of the runners. The team may pick its way across a heavily crevassed glacier with one man roped to the lead trace walking gingerly ahead probing the snow for hidden "slots". Perhaps he will fall through a bridge, to hang on the lip with only dogs and braked sledge holding him, or perhaps he will call an extra lunge from the dogs to pull the sledge from a crumbling bridge while his companion is dragged clear. The course may be across ice so hard and high that the whole sledge bucks and twists like a Viking ship in a seaway. An explorer of Viking stock, Fridjof Nansen, designed the prototype of the sledge last century and had the curved and springy wood bound with rawhide and cord so that it would flex and seem to leap along the ice.

At the end of the long sledging day the Sun will still be high. The dogs are taken off the trace and tied to a wire span anchored in the snow. The pyramid-shaped polar tent is pitched and the "inside man" begins to cook the evening meal while the "outside man" feeds the dogs their frozen blocks of pemmican, carefully keeping the old magazines put inside the pemmican tins as packing, back in New Zealand. Then he cuts heavy snow blocks to anchor the tent against blizzard winds and places smaller blocks between the double walls of the tent at the entrance to be melted for water. He places ration box, medical kit and transceiver radio at the entrance and by the time his tasks are done the inside man is calling "Soup!"

The tent is already warm from the cooking primus. He takes off his mukluks and hangs them to dry at the tent ceiling. Then he takes off his windproofs, puts on slippers of quilted eiderdown and crawls into his double sleeping bag which lies on an air bed down one side of the seven foot square tent. He sits snugly, chews a thickly-buttered sledging biscuit and sips thick, hot soup made from powder. A pressure cooker steams on the primus, boiling stew made from compressed meat bar and dehydrated vegetables later to be served with powdered potatoes cooked by adding boiling water. Probably he is sick of meat bar but there is nothing else until roast chicken and other goodies arrive with a re-supply flight. After the stew he can comfort himself with jam or honey or sardines from the precious box of luxury rations.

Dishes are washed and dried with toilet paper. There is a fog of steam and tobacco smoke in the tent and the primus threatens carbon monoxide poisoning unless air is kept circulating. The men drink tea or coffee, read, play cards, write their diaries. Condensation freezes on the tent's lower walls but with the primus burning the men are quite warm. They do not wash because there is not enough water, but may clean their teeth with damp brushes.

If it is time for the twice-weekly radio call to Scott Base they can talk to "home" tell the world of progress, pass and receive telegrams to and from New Zealand, arrange a meeting with a re-supply or withdrawal aircraft. Morse code often has to be used in bad radio conditions. They go to sleep with the alarm clock ticking

71

among a litter of food bags and utensils on top of the kitchen box which lies between their beds. With the primus out the tent quickly cools but they are warm as they lie listening to the Antarctic silence.

Tomorrow may be a blizzard day, too bad for sledging, when they will lie up in the tent, read the magazines from the pemmican tins, eat light food such as omelettes made from powdered egg, glad of a chance to rest or anxious to be gone, to scale the next peak for a survey station or probe and photograph unusual rock strata on the face of a nearby nunatak.

A dog whimpers and then howls. "That's Kari," they say. "She always starts first." A dog joins her, a low howl deep in a thick-furred throat, rising to the long wavering cry of the wolf ancestor in the Arctic homeland. The "howlo" begins as all the dogs throw back their heads to mourn the night which never comes.

10. *The Machines*

To LIVE IN AN ANTARCTIC BASE IS TO LIVE INSIDE A MACHINE, complex, all its parts carefully integrated, safe. A base is like a ship, a self-contained, mechanical unit with shelter, food, fuel and the engines which motivate each of its parts. The essential engine of the machine is that which produces heat, for to enter the warmth and shelter of an Antarctic hut is like leaving the weather-torn bridge of a ship for the calm below decks.

Because the base is a machine it is noisy. Life inside it continues against a constant, throbbing background of engines producing electric power, pumps feeding fuel to oil heaters, fans whirring to circulate heated air so that it is economically used. If the engines ever stop the silence is oppressive, there is a sense of chill, the whimper of wind in aerial masts and round doorways seems faintly malign. When the noise returns the sense of heat and shelter is immediate and reassuring, but with it comes the reminder that in Antarctica human life continues only under the machine's pledge of safe conduct.

Modern engineering technology has entered every aspect of life in Antarctica. Massive air and sea logistic support of field and base operations would not be possible without icebreakers to crack channels through the sea ice for cargo ships in spring. The pressure cooker enables field party men to cook more and better food with less fuel than was possible fifty years ago. Mixing machines permit ice cream to appear frequently on base menus. Bulldozers prepare air strips to make the operation of large wheeled aircraft possible. Heaters burning kerosene have taken the place of inefficient pot-bellied coal stoves now rusting in the huts of heroic age explorers.

Antarctic engineering has developed as a science in its own right with things to say on subjects as diverse as the efficiency of various types of windproof clothing, the problems of laying concrete building foundations in below-zero temperatures, and methods of navigating vehicles in a straight line in the absence of reliable magnetic compass bearings.

While most machines at work in Antarctica have their counterparts in use in other areas of the world such as the Arctic, the primary Antarctic machine, the complex of huts which make up a base camp, has no parallel. It must be designed to withstand lower temperatures and stronger winds than any other type of building. Men must be able to build it and complete its interior in the shortest possible time during summer. It must survive with little maintenance and it must contain within its walls equipment to fulfil every need of its inhabitants for quite long periods of time. Some buildings must be designed to be buried in snow and withstand the ever-growing pressure of the snow's weight. Others need to be firmly secured to the ground in areas where the permafrost is too hard to permit proper foundations, and to face the weather in such a way that blizzard snow drifts do not bury them.

Designed by New Zealand's Ministry of Works architects, Scott Base is similar to the Antarctic stations manned by several other nations. Its basic concept of a group of huts, self-contained as insulated and heated units and stemming from a central covered way, has proved successful and adaptable.

The base consists of ten connected huts together with a large aircraft hangar now used as a storeroom, and four outlying huts housing scientific instruments. In the connected living-complex are huts containing laboratory and photographic darkroom, mess and galley, cool store and ready-use foodstore, recreation, communication, administration and Post Office facilities, two dormitories, ablutions, carpenters' shop and sledge repair room, electric power generator room engineering workshop and vehicle garage. All these huts are built of pre-fabricated sections drawn together by steel tie rods running through them. Doors are like refrigerator doors and windows small and double-glazed. Before any hut is shipped south it is carefully pre-erected in New Zealand and the parts code-numbered to ensure the shortest possible erection time.

The design and erection of a hut is only part of the problem of providing an efficient living unit. Heating involves a system of cold and hot air inlets and outlets to provide a circulation system which will mitigate the temperature gradient (the rise in temperature from freezing floor to warm ceiling). Each hut has a cold porch like a decompression chamber, with doors on each side. The porch houses the heating unit, usually a thermostatically controlled electric-fired oil burner, and the fan system which pumps hot air into the hut along ceiling vents. Heating units also help to dispose of liquid sewage, either by heating it before it drains down a pipe to the sea ice tide crack, or by blowing hot air down a pipe surrounding the sewer line. Solid refuse is left until it freezes and is then dumped on a rubbish pile in the pressure ridges where it slowly sinks into the sea.

Electricity at Scott Base must be carefully apportioned to various appliances or scientific instruments. The main generator engines are two big diesels which run alternately, and auxiliary engines have to be maintained in case of breakdown. Exhaust gases from the generator motors also heat one of the three base snow melters which are filled with snow every day to provide a water supply. Other melters are operated by oil-fired heaters. Breakdowns or irregularities in the power supply can have drastic consequences for the scientific programme as well as completely disrupting domestic life in the base. Scientific instruments are programmed to automatic recording and valuable records can be lost if there is an undue fluctuation in the current supply.

In winter and bad weather, most of the base staff need not go outside for days at a time because the base is so self-contained. The machines are designed to make living as comfortable as possible. Pumps draw fuel from the fuel tank unless temperatures are so low that the kerosene turns to jelly. In the photographic darkroom where record films from observing instruments are processed, a humidifier puts moisture into the desiccated air to help the film preserve its proper qualities. The galley stove is oil-fired. An electric fire alarm system is ready to register in any hut. Each hut is connected by telephone or signal buzzer which can call a general meeting. The bath water boils in a big copper tank.

Maintenance of the whole system is in the hands of the base engineer, electrician and carpenter who are supported by assistants in the busy days of summer. Their jobs demand continuing vigilance and, often, the most arduous outdoor work when repairs and maintenance are needed for any part of the base and its machinery. They must perform intricate tasks with freezing metal, improvise repairs when no spare parts are available for vital equipment.

Most vehicles used in Antarctica are tried types developed for Artic conditions, but the harsher Antarctic environment with its lower temperatures and more icy travelling surfaces creates new maintenance problems. Transport at Scott Base is provided by several types of vehicle from a Landrover with chain-covered wheels, which can be used on hard snow conditions, to the Nodwell and Snocat which are the largest cargo and personnel-carrying vehicles used. Snotracs, tractors, motor toboggans and small bulldozers complete the fleet. Except for the motor toboggans all operate on the track principle but the tracks must have extra width to distribute ground pressure and prevent the vehicles sinking into soft snow.

Ferguson tractors with their loose metal tracks which fit over conventional wheels of the farm tractor are particularly versatile for New Zealand Antarctic operations, and were initially tried and proved on the Trans-Antarctic Expedition depot-laying journeys from Scott Base to the South Pole. Some of the original vehicles of that journey are still in use at Scott Base along with one of the Snocats driven by the crossing party.

Speed is not a quality of any of these vehicles but on dangerous ice terrain it is, in any case, not desirable. Their chief attribute is power, traction to move in

75

extremely rough conditions, and reliable running in low temperatures. As well as —in the case of Nodwell, Snocat and Snotrac—carrying passengers and small quantities of cargo inside, Antarctic vehicles must be capable of towing sledges. These range from the light, springy motor toboggan sledges, similar to a dog sledge, to the five-ton, steel-shod cargo sledges pulled by Snocat and bulldozer. When ships and aircraft come to McMurdo Sound with cargo for the base, sledge trains may have to transport it across the frozen sea.

Sometimes the sea ice is so dense that cargo must be discharged from a ship moored a dozen miles down the Sound. Its transport to the base becomes a large-scale operation, trying for both men and vehicles and always of uncertain safety. At such times the mechanic who has to change a broken track or replace transmission parts shattered with cold and hard work must display a patience and skill that temperate climate conditions would not require of him. And there can be few colder jobs in Antarctica than to sit in the unprotected driving seat of a tractor towing a sledge train up McMurdo Sound at three miles an hour against the bite of a southeasterly breeze.

The icebreaker has perhaps done more than any other machine to make large-scale Antarctic operations possible. The early explorers took their little wooden ships as far south as the head of McMurdo Sound but their entry was only by chance permission of the ice. They had no guarantee of being able to return. Today United States Navy icebreakers can make channels through fast ice up to fifteen feet thick through which cargo ships with their thin plating can carefully make their way. Each spring the icebreakers of Operation Deep Freeze must break anything up to fifty or sixty miles of channel to allow the first cargo convoy to reach the Sound with fuel, food, vehicles, building materials and scientific stores urgently needed to begin the new summer operation.

With diesel-electric engines which can stop, start and vary the motion of their propellors almost instantly, the icebreakers are supremely manoeuvrable ships. They break ice by ramming its edge with their sharply cut-away bow so that it rides up onto the ice which breaks under the downward pressure. Round-bottomed, they have vast heeling tanks which can be pumped full or emptied of water to create a rolling motion and increase the breaking action. Broken ice is cleared astern by the current flow generated by their propellors.

Though strengthened to withstand the pressure of ice, New Zealand's Antarctic supply ship, H.M.N.Z.S. *Endeavour*, is a little ship compared with the U.S.S. *Glacier* the largest American icebreaker. The *Endeavour* has a fuel-carrying capacity of 1,000 tons as well as dry cargo space, accommodation for Antarctic staff and oceanographers, oceanographic laboratories and a cool store to carry fresh vegetables south for the Scott Base staff. Each summer she makes several trips south to be met at the pack ice edge by an escorting icebreaker. Her operation is

76

part of a growing tradition of Antarctic service for the Royal New Zealand Navy as she is the second Antarctic supply ship to be named *Endeavour*. The first was a vessel of wooden construction bought by New Zealand to support the T.A.E. and I.G.Y. Antarctic operations. The former *Endeavour* spent five seasons voyaging south before being retired.

The Royal New Zealand Air Force, through its Antarctic Flight, also supported Antarctic operations for a number of years with a ski-equipped Otter aircraft and an Auster with both skis and floats. The Otter was lost during an operation in the Trans-Antarctic Range and the Flight was disbanded. The R.N.Z.A.F. role in Antarctica has since been taken over by Hercules transport aircraft which assist similar United States planes in the annual movement of men and materials, although Operation Deepfreeze still provides most air support for operations on the continent.

Radio communication systems are also essential for the conduct of modern Antarctic operations. As the radio centre for New Zealand's Antarctic programme, Scott Base has two functions. It must communicate with the operational planners in New Zealand and with parties in the field. New Zealand-designed field radio sets have been particularly successful in spite of the very difficult conditions under which they must work. The base communications system is maintained and operated by the New Zealand Post Office which provides operators and technicians each year. Communications with New Zealand are maintained on both telegraphy and voice circuits. Vanda Station communicates with New Zealand via Scott Base.

Through the New Zealand Post Office, the Ross Dependency Post Office, as it is called (it is an international Post Office issuing its own Ross Dependency stamps), can link with the telephone subscribers anywhere in the world allowing base staff to talk with families and friends on daily schedules throughout the year. There is no guarantee that radio wave propagation conditions in the ionosphere will allow the link to be established but telegraphic communication is usually possible even if a voice circuit is untenable.

New Zealanders have long been known for their mechanical aptitude—their ability to make things work in bad circumstances and to improvise solutions to mechanical problems when proper facilities are not available. New Zealand jokes about farmers who mend their machines with hay-baling wire and run them for years on end are legendary, and while such rudimentary mechanics are not likely to avail in Antarctica the same human qualities make for the efficient running of machinery. There, machinery must work because lives can depend on it and comfort even more so. Every man who goes to Antarctica must be in some sense a handyman able to think about machines and to make them work for him.

11. *The Men*

COMPLETE WITH DANCING, DRINKING AND A FEW FIGHTS THE first Antarctic party took place some 130 years ago on the pack ice of the Ross Sea. It was held to celebrate the New Year of 1842 by the crews of Sir James Clark Ross's ships, H.M.S. *Erebus* and H.M.S. *Terror*, moored to a large ice floe in latitude 66°32′S.

"Here was a Game in the Antarctic Seas. A public house Erected on the berg with all Kind of Games. A grazy tailed pig. Climbing a grazy pole. Jumping in a bag." With somewhat eccentric spelling and punctuation, the Irish blacksmith on H.M.S. *Erebus* so described the scene in a letter home.

"The Terror's Crew Came on board we Kept up Dancing until 5 o'clock in the morning. When it Ended with three or four Pugilastic matches in the Forecastle which peaceably Ended. All that day the Boatswain and Crews were preparing a ball room clearing away the snow to erect a Public House which was completed by noon adjoining the Bar of the Tavern there was a circus for Different Kind of Games.... The sign of the Public House was The Pilgrim of the Ocean. At the reverse Side of the Board The Pioneers of Science.... The Games went off well the Exhibition in the circus far Exceeded the Waltzing in the Ball room. James Savage carried the prize in the Bag. Jatter Welsh half strangled the pig and Bandy Carried the prize for the pole. When the Essence of the Barley heated our Gents the Snow Balls went flying." *

Apart from the period sports it was a party that might have taken place at any modern Antarctic base. Quite obviously it was a good party. It dissipated the tensions, fears and aggressions of a small group of men living together in very strange surroundings far from home. It was an opportunity to let off steam given by an expedition leader enlightened for his day. It was all very pleasant and natural.

Among men chosen for the physical fitness and general competence to live and work in Antarctica, and who are provided with the best possible facilities for

* From a MS published in *Polar Record* Vol 10 No. 69. Reprinted by permission of the Scot* Research Institute, Cambridge, England.

maintaining their health and comfort, the biggest problems are those of orientation to a new environment and to an unvarying group of companions.

Both these factors have immediate and drastic effects on emotional and nervous systems, effects which must be dealt with by the individual because there is usually no way of going back to the comfort of home and familiar things. Now that largescale operations are being continued by several nations, all with differing approaches to physical and mental problems of adjustment to severe environments, and with different national characteristics reflected in their expedition members, Antarctica has become one of the world's most important social and medical laboratories.

The Antarctic environment has no parallel, for even the Arctic has the tundra, its stunted vegetation, its variety of animals and relative accessibility. To travel from spring in New Zealand to spring at McMurdo Sound is to undergo an environmental temperature change of up to 150°F. The Antarctic world is black and white without the usual visual signposts of colour. The Antarctic light and dark are twenty-four hours long and contradict the normal diurnal rhythm of the body's responses. Cold decreases the sense of touch, heavy clothing makes movement ponderous, the blizzard wind is both blinding and deafening. The extreme dryness of Antarctic air aggravates the dehydration which is part of the body's response to cold. All these things greatly influence the reactions of men in Antarctica to the world about them and to their fellows. Psychological adaptation is affected by physiological adaptation and physical responses in turn reflect mental conditions and attitudes.

Indoors, however, men can now provide themselves with their preferred climate, food, colour schemes, music, literature, and recreation facilities. Antarctic life indoors has little in common with that in the unheated huts of the polar heroes. This is another factor to which newcomers to Antarctica must adjust. The exploits of the pioneers were such that people still imagine Antarctic life in the old terms of primitive living, the desperate struggle for existence. To realise that he is not a hero after all is a crushing feeling for the romantic man who goes south and finds that extreme demands are not normally to be made on him.

With these considerations to be kept in mind the selection of men for Antarctic work has some hazards. Even if expedition leaders and organisers are not prepared to rely entirely on their personal judgment of men's characters to make satisfactory selections, scientific psychological testing must be carried out in the face of many imponderable factors in the Antarctic environment. Tests on which selection decisions can be made with certainty have not been devised. Finally the question is simply "Is this man fit for life in a small, cramped community?" It must be answered in terms of an overall assessment of the man's usefulness to such a community rather than on his apparent personal merits.

Antarctic sociological research is progressing slowly and in deeply concerned with certain problems—the psychological effects of monotony and ways of alleviating it; seasonal fluctuations in mental states and their physical manifestations; methods of giving prospective personnel accurate information on the conditions they will meet; the incidence of insomnia, particularly during the Antarctic winter; reaction to sexual deprivation; and the study of the individual's needs for personal privacy. Such researches are concerned with ways of measuring the individual's adaptation to Antarctic life as a continuing process rather than describing the qualities of the adapted individual.

Many other problems are now recognised and new ones continue to be defined. It is suggested, for instance, that among small groups better feelings of efficiency and achievement result from a democratic type of base leadership, while with larger groups the opposite is the case. In an industrial world concerned with production efficiency, when large expenditures of money and materials are being made for Antarctic research, the style of leadership adopted by a base leader faced with these alternative methods of organising his men assumes great importance to expedition economy.

Research has already shown that food is the most abiding interest of groups of men living in Antarctic isolation. Variety in food supplies is well recognised for its monotony-breaking action, but medical and nutritional researches are still needed to indicate the most appropriate eating habits for men in cold climates. In spite of the fact that a higher food intake than for temperate climate is necessary to supply muscular energy and maintain body heat in Antarctic cold over-eating, particularly during the inactive winter, can cause problems.

Cold climate nutritional research is directed towards discovering the most satisfactory type of balanced diet. There is evidence that while protein foods give the body energy for a short-term temperature rise, fats and carbohydrates do more to sustain a high heat production. The role of vitamins in keeping the body healthy in polar conditions is not fully understood. It is not known why the fresh seal meat diets adopted by early explorers to ward off the vitamin C deficiency disease, scurvy, should be so successful when a meat diet does not contain as much vitamin C as the recognised standard for temperate climate conditions.

Dietary problems are related to human cold acclimatisation and so to the whole field of Antarctic medicine. The physical capabilities of the human body in extreme cold depend ultimately on the maintenance of heat balance in the body. This, in turn, depends on the degree of cold, the production of heat by food converted into expended energy, and the conservation of heat by clothing. These capabilities are not well known and research is needed to find out the nutritional requirements of individuals in different Antarctic occupations. Other physiological research into the processes which maintain heat balance are complementary to nutritional studies.

While problems of physiological adaptation to Antarctic conditions appear to be of more immediate importance medically than those related to the cause and occurrence of disease, research work is still necessary in medical microbiology. The existence of Antarctica as a simplified microbiological environment means that the behaviour of particular types of disease-causing microbes and the way they affect men can be easily studied.

Men working in isolated groups in polar regions are usually found to be remarkably fit and free from infections. Respiratory system bacteria, for instance, have been found to decrease in individual concentration during periods of Antarctic isolation. Particular strains of bacteria causing infection can be easily identified in the simplified environment, and their natural history studied in a manner impossible in temperate areas where vast numbers and varieties of bacteria are circulating among large human populations. Another aspect of Antarctic disease is the discovery and description of microbial life carried by Antarctic animals and harmful to man. This question is also directly concerned with the estimation of man's influence on such animals, a task being undertaken with some sense of urgency because the degree of man's contamination of Antarctica could become so great that it obscures the nature of the original life.

In planning Antarctic expeditions all these problems of environmental adaptation, selection of appropriate personnel, and provision of the most satisfactory base living conditions must be solved on a trial and error basis, with the assistance of the increasing body of polar knowledge. New Zealand entered Antarctica having designed and prepared Scott Base on the basis of the experience and knowledge of other national expeditions and of New Zealanders who made preliminary examinations of conditions at McMurdo Sound. The result is a compact base entirely adequate for its scientific and logistic purpose; and a valuable collection of experience on all Antarctic problems, from the successful design of windproof clothing to the techniques of dog handling.

Scott Base today is more a hostel for scientists rather than a refuge for explorers. Newcomers are surprised at the spaciousness of the mess room, lounge and recreation room, the walls lined with books, the shelf of gramophone records, the variety of food. A bath in its little bathroom complete with weight scales, a washing machine in the ablutions room, and sewing machines in the sledge repair room are surprising in their domesticity. The red telephone in its carefully sound-proofed telephone booth beside the Post Office is just like those in any New Zealand home into which its user talks.

Except when field parties and visitors increase the numbers living at the base each man can find a private haven in his sleeping cubicle with its two bunks, chair and writing desk. The hospital with its surgical equipment and rows of drug bottles could be anywhere, and its presence is as re-assuring as the knowledge that full medical and dental facilities are available to New Zealanders at McMurdo Station.

Working hours at Scott Base are set by individual leaders but the tradition is to work six days a week with Sundays off. Work makes for high morale in the

SOUTH

Antarctic and the forty-hour week seems pointless when there are jobs to be done without the diversions and distractions of life at home. The day begins at seven a.m. and ends about eight p.m. In the evenings there may be films or radio telephone schedules with New Zealand. Some men like to go skiing on the base ski field a few hundred yards away. It has its own tow run by a World War II bren-gun carrier abandoned because its narrow tracks proved unsuitable for Antarctic snow travel.

It is as well that a work shedule is adhered to for time does not have as much meaning in Antarctica as at home. There is always the inclination to remain awake at night absorbed in some occupation, and to "cat-nap" during the day, with a consequent disorganisation of routine which does not help the general tenor of base life. In a small, closed society it is better for all if all act similarly although tolerance must continually be exercised. A must be prepared to let B play his favourite gramophone record once in a while, even though A violently dislikes it. All members of the party must be ready to do outdoor work in unpleasant conditions for dissension is caused if anyone apparently shirks his duties.

In spring and autumn when regular flights are being made between New Zealand and McMurdo Sound there is plenty of opportunity to send and receive letters. The expected arrival of a mail-carrying flight is accompanied by general tension. Hometown newspapers are coveted as reminders of things loved and missed. In winter mail is, of course, not available and the group must discover other emotional outlets. Deep and lasting friendships are created. If you winter over with a man you know him very well.

In the messroom at Scott Base is a sideboard. In winter it usually bears bottles of liquor left over from Saturday night and slowly emptied with after-dinner conversations during the week. It also bears a bottle of vitamin tablets, jugs of fruit juice and ingredients for making hot beverages such as cocoa, because consumption of liquids is high in the dry climate. In Antarctica these things are not luxuries but necessary dietary additions and compensations for the disadvantages of isolation. In the same way the hundreds of pounds of chicken, turkey, goose, fillet steak, ham, crayfish, oysters, exotic canned fruits, nuts, cheeses and other delicacies eaten during a Scott Base year are essential items for making life a little more interesting and satisfying.

Much care goes into the ordering of food for each year's base resupply, and just as careful an account is taken of the expedition members' preferences in pipe and cigarette tobacco, soft drinks, music, chocolate, soap, toothpaste and even shaving cream for, strangely enough, not everybody wants to grow a beard. The Scott Base store rooms are wonderlands of consumer goods all of which are vitally necessary for the maintenance of health and morale. The meat-cave cut deep into a snowfield on Crater Hill and spangled with ice crystals is a miraculous butcher's shop from which the cook daily draws cuts to refresh jaded palates.

82

The Scott Base Leader is entirely responsible for all that goes on in the base as well as for the New Zealand personnel sharing Cape Hallett Station with Americans, even though the general control of Hallett is exercised by an American naval officer. The Leader is a man of wide experience in similar environments, often with a background of service in the armed forces. He is responsible for the successful completion of the year's research programme as well as the upkeep and good management of the base. He is also the channel through which liaison with Operation Deep Freeze is carried out in Antarctica. His tasks are complex and varied as he must supervise field operations as well as the preparation of materials and equipment for each year's base resupply. He must know something of science to be able to support the base scientific programme, something about engineering to be able to make decisions on the operation of mechanical equipment, something about veterinary science, something about accounting; and a great deal about men.

The Scott Base leader also has official duties as the Queen's representative in the Ross Dependency, New Zealand's Antarctic territorial claim. Under New Zealand law he is a Justice of the Peace, a Coroner and a Postmaster and he must know something about all these fields of administration.

New Zealanders who go south—be they surveyors, carpenters, mechanics, cooks or geophysicists—are inclined to be quiet, self-reliant people with their own quick brand of national humour which sometimes baffles American visitors. Their qualities are reflected in the way they look after their dogs, mend their sledges, operate their vehicles, patiently run their scientific instruments and revere the photographs of their Queen and of Captain Robert Falcon Scott on the messroom wall. They form close-knit groups, hospitable to strangers, not over-curious about what other people are doing, but intensely interested about the broad fields of human endeavour, and particularly, about exploration and life in extreme circumstances. In a way, they feel they are at home in Antarctica.

12. *The Future*

ON THE SHORES OF ROSS ISLAND AND VICTORIA LAND THE HUTS of British expeditions which pioneered Antarctic exploration still stand. They are regarded as historic monuments by men of all nations who visit the Ross Sea region and contribute to New Zealand's strong sense of an inheritance in Antarctica only lately taken up yet highly valued. As a nation founded by England, New Zealanders feel they are custodians of the things which England left in Victoria Land and of the traditions of exploration and Antarctic living which the English expeditions established from bases at Cape Adare, Hut Point, Cape Royds, and Cape Evans.

In the summer of 1960–61 the New Zealand Government sponsored a project to restore Scott's last expedition hut at Cape Evans and the Shackleton 1907–09 expedition hut at Cape Royds to their original appearance, and to preserve them from further damage. The New Zealand Antarctic Society, a national body with a large public membership, provided three men to do the work. Antarctic Division arranged for its execution on the basis of a joint decision to which the New Zealand Historic Places Trust, the Royal New Zealand Navy, the Ministry of Works, the Ross Dependency Research Committee and the Dominion Museum also contributed. Now maintained in a condition similar to that in which they were left, the huts bear bronze plaques recording the expeditions which used them and framed scrolls presented by the Society briefly telling their history.

The hut at Hut Point, used by Scott's *Discovery* expedition of 1901–04, for many years filled with ice from drift snow accumulation, was excavated and permanently secured against the weather by a Society working party in the 1963–64 summer. At Cape Adare the hut of Borchgrevink's *Southern Cross* expedition in 1899, the first to winter on the continent, still stands. Beside it are the ruins of the hut used by the northern party of Scott's last expedition. At Inexpressible Island,

84

100 miles south of Cape Hallett, lie the remains of the cave in which the party later spent an incredible winter of hardship. At Granite Harbour, further south, is the remains of the summer rock shelter built by the western geological party of Scott's last expedition. At Cape Crozier are the remains of the stone and canvas igloo built by Dr Edward Wilson's party of three who made their valiant winter journey in search of Emperor penguin eggs in 1911.

The hills above the base huts of several expeditions are surmounted by crosses which honour and commemorate men who died in Antarctic service—Scott and his polar party on Observation Hill; the seaman, Vince, on Hut Point; Spencer-Smith, Mackintosh and Haywood from the Ross Sea party of Shackleton's tragic *Endurance* expedition on Cape Evans; the biologist Hanson of Borchgrevink's party above his grave on the heights of Cape Adare. Beside the flagstaff at Scott Base stands a simple tablet to the memory of a New Zealander killed in the field in 1961. The name of Williams Field, the McMurdo Sound airport, commemorates an American serviceman; and another is honoured on the hill above McMurdo Station.

Exploration will always have its cost in life but as Antarctica becomes better known and understood its dangers lessen. While by its very nature it will always be a land which threatens life, emotional attitudes to it are rightly changing. If they did not they would, in a romantic glow of awe and hero-worship, obscure the real place which Antarctica now holds in the lives of people in all parts of the world as a great source of new knowledge and understanding. Antarctica's geographical isolation; its vastness; its cold and wind; its great heights; its covering of a foreign element, ice; its lack of native human population; these matters must no longer cause a public blindness to its importance both to scientific research and to politics in a world struggling towards international government. In the modern world these very features have become virtues rather than disadvantages in putting Antarctica to work for mankind.

The Antarctic Treaty Conference and the Scientific Committee for Antarctic Research, which, because it is a body of the International Council of Scientific Unions (I.C.S.U.) reflects the internationalism of the Treaty, continue to function with outstanding success. By its system of breaking down its discussions into working group meetings such as those on solid earth physics, upper atmosphere physics, oceanography, biology, geology, geodesy and cartography, and logistics, each comprised of national representatives and chaired by scientists from different nations, S.C.A.R. can achieve a notable exchange of knowledge and systematic research planning.

The periodic symposia held by working groups give an opportunity for large-scale exchanges of current research findings and are supplemented by the work of other bodies such as the World Health Organisation with its 1962 conference on medicine and public health in the Arctic and Antarctic. Through other organs of the I.C.S.U., such as the Committee on Space Research, Antarctic science is also

85

furthered. A case in this instance is the use of Earth satellite measurements to study the distribution of pack ice on a continental scale or accurately position the Antarctic coastline to record changes in the shape of the ice sheet.

No accurate forecast of Antarctica's future for mankind can yet be made but many changes in both research emphasis and in the organisation of daily living are bound to come about as science and technology progress. Upper atmosphere studies may well be reduced in intensity with the end of the present sunspot cycle and greater emphasis given to Earth and life sciences. Terrestrial and marine biology will continue to be emphasised as national research programmes develop. This will mean that the costly and complex ice sheet stations of the Antarctic interior will be reduced in size and population or even transformed into automatic recording stations as scientific equipment becomes more sophisticated. Population will concentrate in the coastal areas. The question of whether women are to be included in national expeditions will be answered in due course but at present winter station life remains for good social and economic reasons, a male affair.

As the Southern Hemisphere population increases, the world's civilian airline services may demand airports on the continent to service trans-polar air routes like those across the Arctic. World shortages of minerals may demand the working of economic Antarctic deposits which may yet be found. Some exploitation of Antarctic marine life is bound to come about, although this would not necessarily need bases or processing plants on the continental shore. A tourist industry is no longer a visionary's dream: tourist ships are already visiting both McMurdo Sound and the Antarctic Peninsula. It is agreed that tourist visits should be strictly controlled to prevent disruption of wildlife and continuation of the "clean" Antarctic environment as well as avoiding hazard to the tourists or the scientists and support staff who may become needlessly involved in rescue operations. Even though technological advances in transport and polar living have at last put the continent within easy reach it remains a wild and eternally hostile land where the overconfident traveller will always be in danger. The chief field of activity is likely to remain squarely on its most successful present basis—the prosecution of the scientific research which has already proved so valuable.

Like the mountains of his Southern Alps, Antarctica offers to the New Zealander a chance for great adventure. He adapts well to polar life but anticipates the joy of his return home to a green and peaceful land with sentimental though seldom expressed intensity. Most likely he would echo the words of the blacksmith on H.M.S. *Erebus* who wrote of his return to port "All hands were in Good health and Spirits, Fresh grub, Liberty on Shore with a drop of the Creator—Soon made our Jolly Tars forget the Cold fingers in the Frozen Regions...for very Little they thought of 78 South while Regealing them Selves at Charley Probins the Sign of the Gordon."

THE PLATES

THE PHOTOGRAPHERS

Guy Mannering 2, 3, 7, 13–15, 18, 23, 33–35, 37, 39–41, 43–45, 47, 51–53, 55–57, 60, 61, 64, 65, 78–80, 96, 97, 102, 103, 105, 107–109, 117, 118, 123–125, 128, 129, 137, 139–142, 144, 149–151, 158, 163–167, 169, 171, 172, 179, 181, 183–186, 192, 193, 197–201, 204–206; H.D. O'Kane 5, 48, 54, 71, 81–83, 86–88, 94, 143, 156, 173, 174, 180, 182; P.M. Otway 1, 16, 27, 73, 84, 90–93, 99, 145, 160, 175, 196, 203; R.S. Cranfield 21, 30 67, 69, 170, 187, 189, 202; B. Reid 10, 29, 66, 115, 119, 130, 131, 207; S. Watts 20, 28, 106, 113, 114, 116, 190; J. Cranfield 11, 120, 126, 133, 135, 136, 138; J. Ricker 12, 24, 68, 147, 148; B. Sandford 38, 98, 100, 101, 188; W.R. Logie 4, 26, 74, 153, 154; R. Shanahan 72, 95, 146; M. Woodgyer 75, 134; C. Bailey 9, 31; N. Cooper 6, 155; G. Warren 25, 159; R. Dibble 32, 36; Trans-Antarctic Expedition Collection 46, 89; A.J. Heine 49, 104; W. Prebble 50, 58; J. Kennett 122, 195; W. Smith 62, 191; H. Gair 63, 161; G. Midwinter 70, 77; N.Z. Oceanographic Institute Collection 110–112; G. W. Grindley 59; R. Henderson 121; C. Taylor 194; U. McGregor 162; B. Woods 157; W. Webb 42, 168; G. Connell 19, 22, 177, 178; John Whalan 8, 17, 75, 76. 167

you walk where
life and non-life
physical movement and motionlessness
are in delicate balance
and yet the land appears relaxed and still.
Here are the world's lowest temperatures
and strongest winds,
hardiest animals
and seas richest in life.

ANTARCTICA
&
THE ELEMENTS

ANTARCTICA AND THE ELEMENTS

3

4

5

6

7

8

9

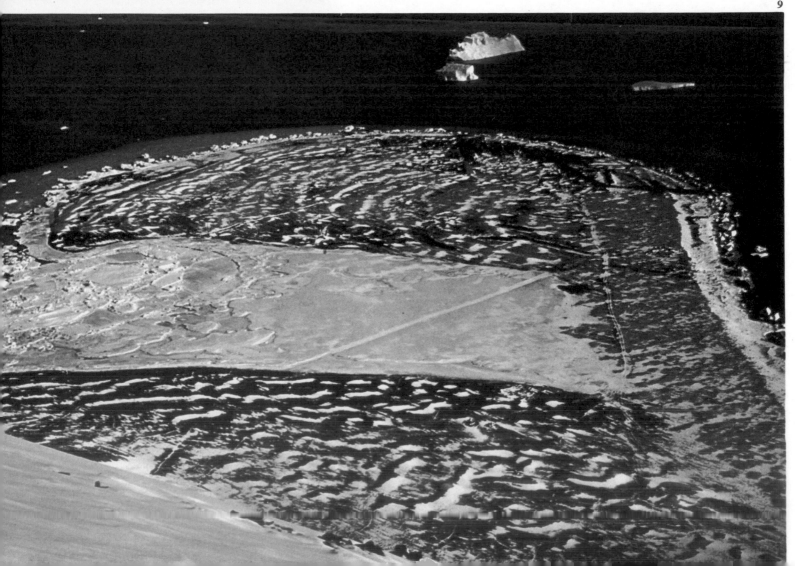

12

13

1

18

19

20

21

22

23

24

25

26

27

28

30

31

like the sea
the ice gives no second chance
but is entirely neutral
in the struggle for survival.
Ice dominates human life
and human sight
which is endlessly amazed
by the brilliance of its surfaces,
the violet, emerald, and cobalt depths
of its shadowed faces.

THE ICE
&
THE LAND

32

34

35

36

37

38

41

43

42

46

44

45

48

51

52

53

57

58

59

62

63

in the black water
you see the ice begin
with the iridescence of its
first minute crystals.
Towards mid-March the last ship
must leave or be trapped
in the freezing sea.
The last farewells are said at dusk.

THE DAY
&
THE NIGHT

64

66

67

68

69

70

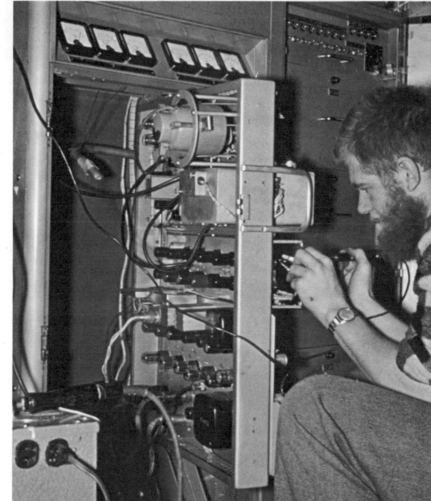

71

72

73

74

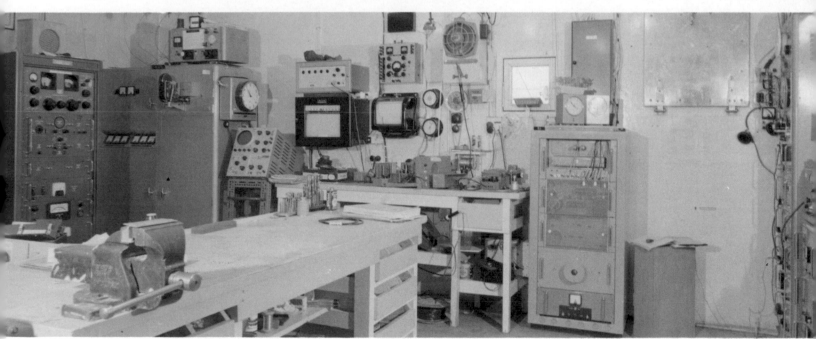

77

79

78

80

83

84

85

86

87

90

91

92

93

94

95

98

99

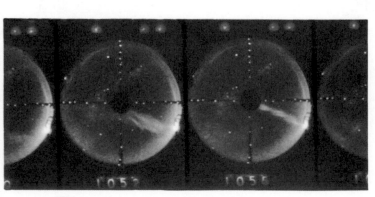

100

101

through these seas
where fog swirls thickly,
where icebergs drift and crumble
and the giant birds wheel
endlessly
on their sickle pinions,
Antarctica stretches a cold hand
to the shores
of India and Arabia.

THE SEA
&
THE ANIMALS

102

149

105

106

107

108

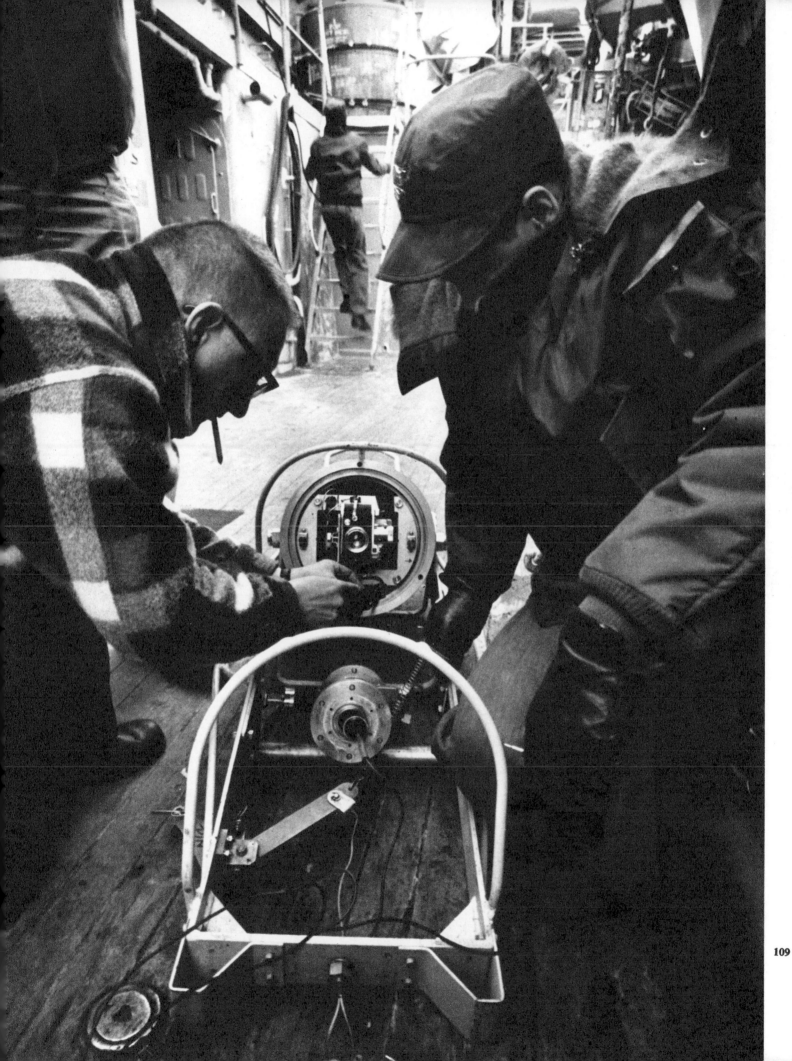

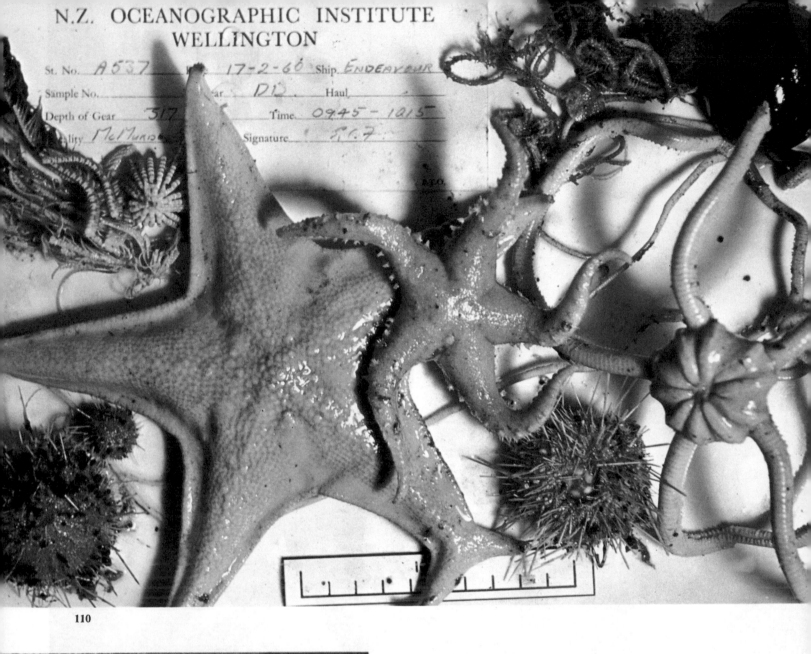

110

111

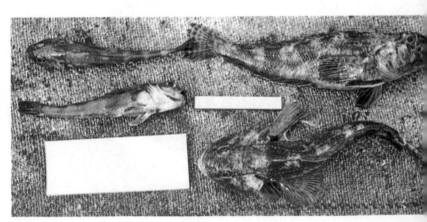

112

113

114

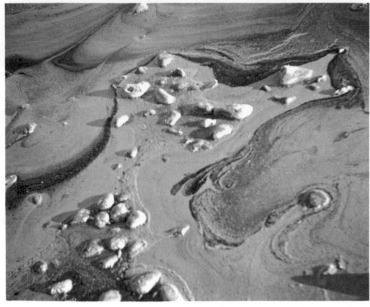

115

116

117

118 ▷

119

120

121

122

123

124

125

127

128

129

130

131

132

133

134

135

136

137

138

140

141

142

on the Plateau
there is no gauge for time
but hunger and cold in the bones.
Distance seems as meaningless
as a dream of home
and the measurement of miles
by revolutions of a bicycle wheel
behind the sledge
pointlessly finite.

THE JOURNEYS
&
THE MACHINES

143

144

145

146

149

152

153

154

155

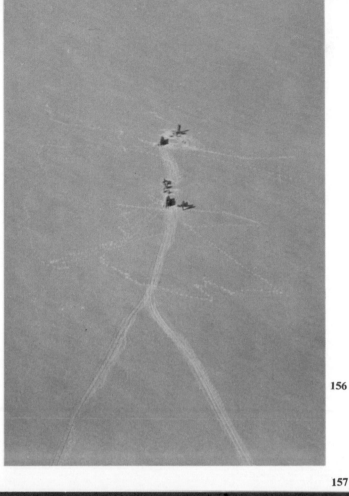

156

157

159

160

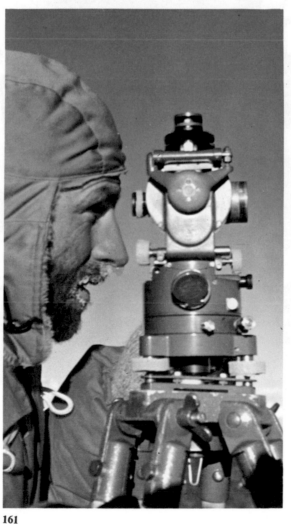

161

162

164

165

166

167

168

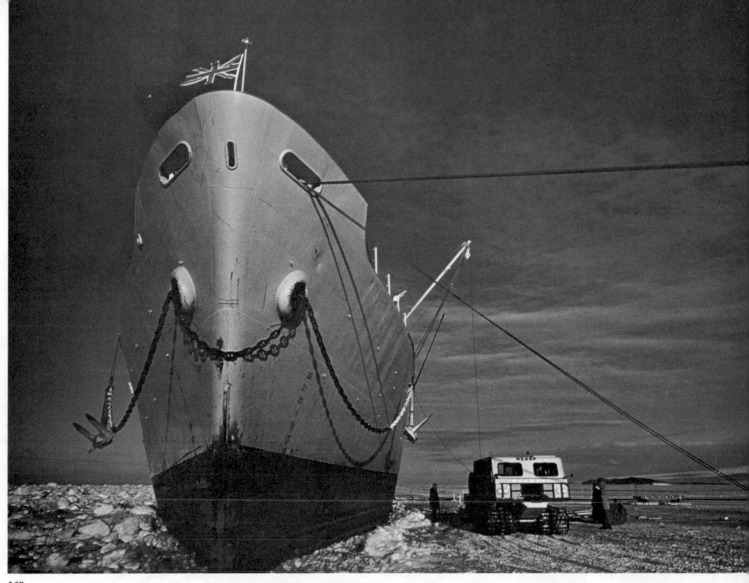

169

170

171

172

173

174

175

176

177

Antarctica's geographical isolation,
its covering of a foreign element,
ice,
its lack of native human population,
must no longer cause a public blindness
to its importance
both to scientific research and to politics
in a world struggling towards
international government.

THE MEN
&
THE FUTURE

178

179 D

180

181

182

185

186

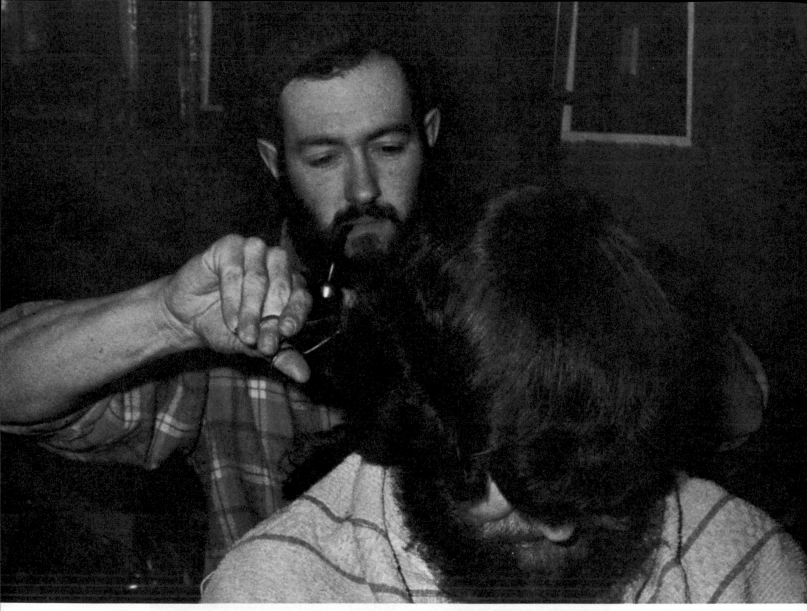

187

188

189

190

191

194

195

196

197

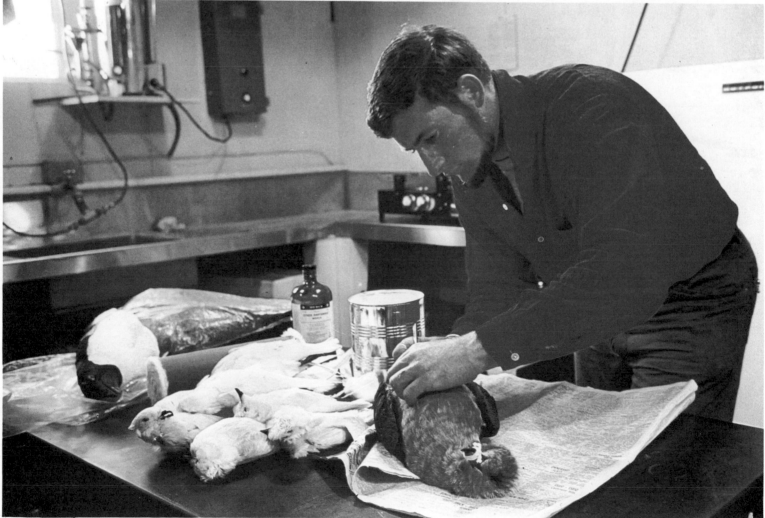

198

199

200

201

202

203

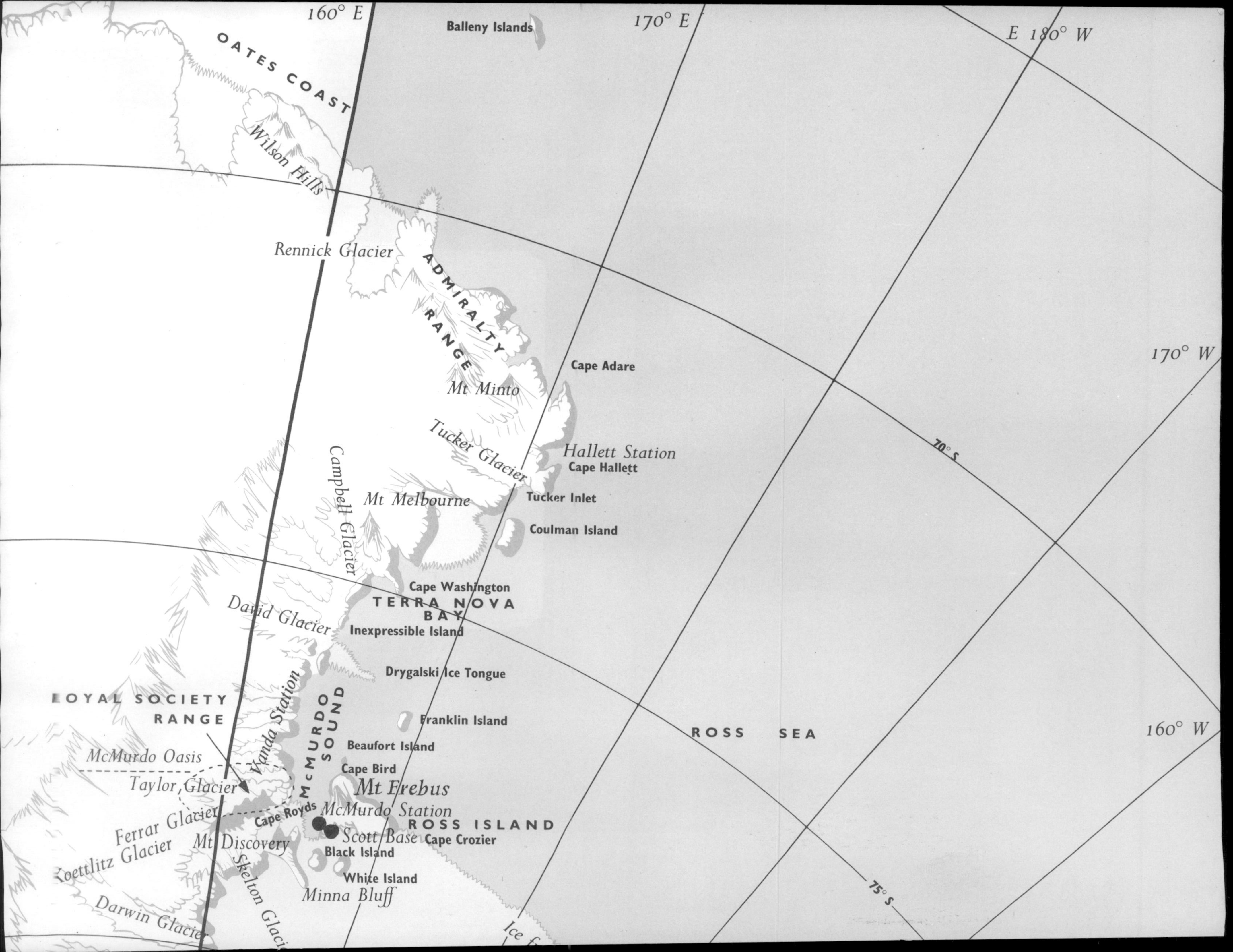
160° E
170° E
E 180° W
170° W
160° W
70° S
75° S
Balleny Islands
OATES COAST
Wilson Hills
Rennick Glacier
ADMIRALTY RANGE
Mt Minto
Cape Adare
Tucker Glacier
Hallett Station
Cape Hallett
Tucker Inlet
Mt Melbourne
Campbell Glacier
Coulman Island
Cape Washington
TERRA NOVA BAY
David Glacier
Inexpressible Island
Drygalski Ice Tongue
Vanda Station
McMURDO SOUND
Franklin Island
Beaufort Island
Cape Bird
Mt Erebus
McMurdo Station
ROSS ISLAND
Cape Royds
Scott Base
Cape Crozier
Black Island
White Island
Minna Bluff
McMurdo Oasis
Taylor Glacier
Ferrar Glacier
Mt Discovery
Koettlitz Glacier
Skelton Glaci
Darwin Glacier
ROSS SEA
Ice f